Hybrid Ship Hulls, Engineering Design Rationales

Hybrid Ship Hulls, Engineering Design Rationales

Editor

Alfred Eckstein

scitus
academics

Hybrid Ship Hulls, Engineering Design Rationales
Edited by **Alfred Eckstein**

ISBN: 978-1-68117-456-3
Library of Congress Control Number: 2016937626

SCITUS Academics LLC,
www.scitusacademics.com
Box No. 4766, 616 Corporate Way,
Suite 2, Valley Cottage,
NY 10989

Printed in United States of America on Acid Free Paper

Preface

A hull is the watertight body of a ship or boat. Above the hull is the superstructure and/or deckhouse, where present. The line where the hull meets the water surface is called the waterline. The structure of the hull varies depending on the vessel type. In a typical modern steel ship, the structure consists of watertight and non-tight decks, major transverse and watertight members called bulkheads, intermediate members such as girders, stringers and webs, and minor members called ordinary transverse frames, frames, or longitudinals, depending on the structural arrangement. The shape of the hull is entirely dependent upon the needs of the design. Shapes range from a nearly perfect box in the case of scow barges, to a needle-sharp surface of revolution in the case of a racing multihull sailboat. The shape is chosen to strike a balance between cost, hydrostatic considerations (accommodation, load carrying and stability), hydrodynamics (speed, power requirements, and motion and behavior in a seaway) and special considerations for the ship's role, such as the rounded bow of an icebreaker or the flat bottom of a landing craft. The book Hybrid Ship Hulls, Engineering Design Rationales provides an overview of cutting-edge developments in hybrid composite-metal marine ship hulls, covering the critical differences in material processing and structural behavior that must be taken into account to maximise benefits and performance. Supporting the design of effective hybrid hulls through proper consideration of the benefits and challenges inherent to heterogenic structures, the book covers specific details of quality control, manufacturing, mechanical and thermal stress, and other behavioral aspects that need to be treated differently when engineering hybrid ship hulls

Table of Contents

CHAPTER 1

Experimental fatigue and aging evaluation of the composite patch repair of a metallic ship hull

Luiz CM Meniconi*, Luiz DM Lana and Sergio RK Morikawa

Petrobras Research Center (Cenpes), Rio de Janeiro, Brazil

ABSTRACT

This article describes the fatigue analysis of a composite repair that was applied to the metallic hull of a Floating, Storage and Offloading (FSO) platform. The main objective is to address the durability and thus the expected operational life of the repair, with emphasis on the adhesive bonded interface between metal and composite. The adoption of this repair technology is increasing in Brazil and abroad and little is known about its long term performance when applied to harsh, dynamic applications like naval structures in operation. During repair installation, more than a year ago, an array of Bragg grating extensometers was applied for reliable structural behavior monitoring. Dynamic strain samples were acquired daily and remotely sent to shore for processing. In parallel, lap shear fatigue tests were performed at the lab in order to establish a suitable defect growth fatigue curve, concerning repair disbondment. The experimental strain data, together with a specific fatigue curved experimentally defined provided the input of a Finite Element Model of the repaired structure and resulted in the expected fatigue life of the repair metal-composite interface. Environmental aging was beneficial as it resulted in a12% increase in the critical shear stress of the interface.

KEYWORDS

Composite repair; Adhesive interface; Fatigue analysis; Aging; Durability; Structural monitoring

1. BACKGROUND

Offshore production structures like Floating, Production, Storage and Offloading vessels (FPSOs) are designed to remain in station for 25 years or more. This is a major deviation from the traditional ship maintenance scheme, which involves dry docking every 5 years or so, for overhaul maintenance. Due to this scenario, in place repair techniques were investigated, in order to restore structural integrity without the need of interrupting production. Composite patch repairs are one of those techniques, because no hot work is involved, turning the operation intrinsically safe. Many success application cases of this technique are reported [1].

In this study case the composite patch repair design followed the approach proposed at the DNV technical report "Project Recommended Practice of Composite Patch Repair for FPSO Structures" (DNV RP) [2]-[4]. The design has also utilized the Finite Element Method (FEM), starting from the global model of the ship as available in a database. The local model of the repaired region was constrained at its boundaries by the displacements obtained from the global model for an extreme load case. The local model of the repair was implemented in ABAQUS™ FEM code version 6.13 [5]. Dynamic lap shear tests were run in order to define a fatigue curve for this specific case and this was compared with strain data acquired at the repaired structure.

2. METHODS

2.1 Double lap shear tests

The composite material adopted for the repair is a biaxial 45°/-45°, non-crimp carbon fabric, as the main objective was to reinstate the shear stiffness of the hull. The resin used for lamination was a rubber modified vinyl-ester. The resin was also applied as the adhesive at the metal-composite interface. The elastic properties of the carbon laminate were, knowing that directions 1/2 correspond to 0°/90°: $E_1 = E_2 = 46$GPa, $\nu_{12} = 0.05$ and $G_{12} = 3$ GPa. The adhesive properties were $E = 2.3$ GPa and $\nu = 0.38$.

A metallic plate, 8 mm thick, measuring 500x650 mm was laminated at one side using the same scheme proposed for the repair, i.e., a first layer of resin, one layer of glass chopped strand mat (CSM), followed by the 45°/-45° carbon lamination. Figure 1 shows a lap shear specimen. The laminated region had dimensions of 370×650 mm.

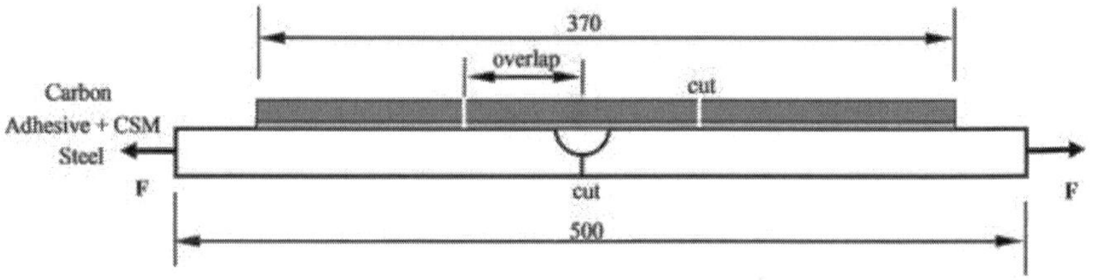

Figure 1. Typical single lap shear test specimen.

Thirteen layers of carbon were deployed, making a total thickness of 8 mm. After cure the plate was cut in strips 25 mm wide. Further cuts were done, both to separate the metallic halves and to define different overlap lengths. Bonding interface properties were evaluated as proposed in section 8.F of DNV RP, with some modifications as described below. The evaluation is based on ASTM D3528 double lap shear (DLS) test method [6], with the difference that single lap shear specimens were tested in pairs, to make it easier and faster to assembly the test plate in the field. A groove, filled with paste, was introduced in the metallic plate to mimic the thickness loss. Table 1 shows test results.

Table 1. DLS test results

Specimen	Overlap length (mm)	Total width (mm)	Failure load (N)	Unit failure load f (N/mm)
40_1	40	48,7	27607	567
40_2	40	49,6	28783	580
40_3	40	49,9	27149	544
40_4	40	48,1	26467	550
60_1	60	47,4	29177	616
60_2	60	48,0	32482	677
60_3	60	50,7	31008	612
60_4	60	49,5	34371	694
80_1	80	49,8	35618	715
80_2	80	49,1	30233	616
80_3	80	53,2	36861	693
80_4	80	48,1	34774	723
185_1	185	48,7	37130	762
185_2	185	50,7	36666	723
185_3	185	49,9	35547	712
185_4	185	48,3	36111	748

The test results are displayed in terms of unitary failure load against overlap length, as shown in Figure 2. From the graph and according to DNV RP it is possible to define two parameters of the adhesive interface system tested, the first one being a maximum effective overlap length of around 100 mm, beyond which the unit failure load reaches a plateau value of 736 N/mm. The latter figure divided by the former defines the critical shear stress of the system, which results to be 7.4 MPa.

2.2 FEM simulation of the DLS tests

The DLS tests were simulated in order to define the FEM interface parameters. Both steel strip and carbon laminate were represented by 4-noded shell elements. The 0.6 mm thick adhesive layer between the two materials was described by the cohesive interface interaction behavior available in ABAQUS, which is a very efficient approach, as the adhesive layer itself does not need to be represented by a finite element mesh, so metal and composite meshes are independent. Concerning interface properties, stiffness is firstly defined, either for normal (K_{nn}) as for sliding and tearing shear directions (K_{ss} and K_{tt}). The interface properties were obtained from the elastic properties of the adhesive:

$$
\begin{aligned}
G_{adh} &= \frac{E_{adh}}{2.(1+\nu_{adh})} = \frac{2.3}{2.(1+0.38)} = 0.83\ GPa \\
K_{nn} &= \frac{E_{adh}}{t_{adh}} = \frac{2.3}{0.6} = 3.8\ GPa/mm \\
K_{ss} = K_{tt} &= \frac{G_{adh}}{t_{adh}} = \frac{0.83}{0.6} = 1.4\ GPa/mm
\end{aligned} \qquad (1)
$$

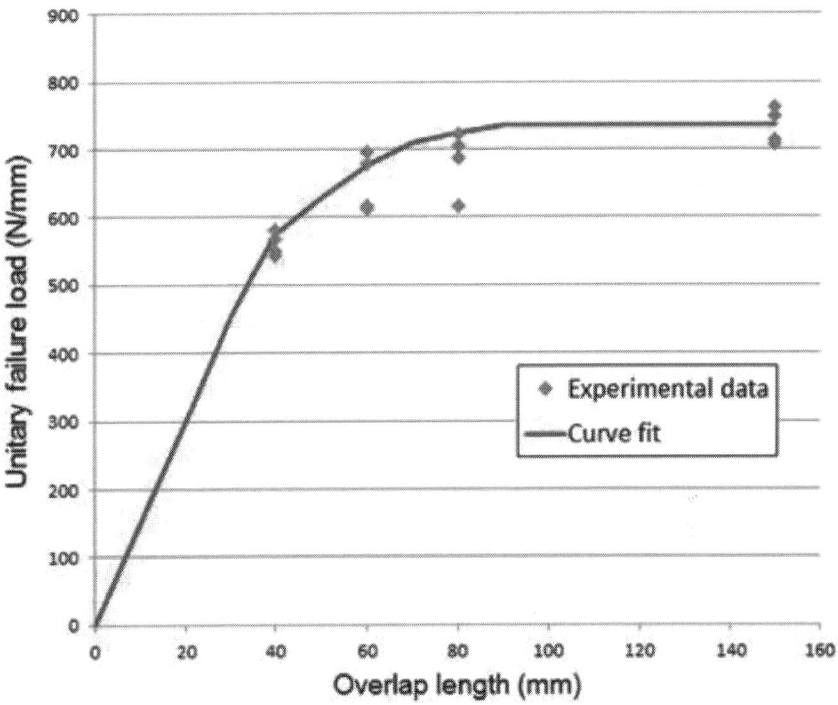

Figure 2. DLS test results: maximum unit failure load.

Next, an interface damage criterion needs to be established. For this, the maximum quadratic stress approach was adopted. The limit normal stress σ_{lim} was considered to be 20 MPa. This can be an arbitrary high value, as measures are taken to reduce the peel stresses at the borders of the laminate. The limit shear stresses in both directions *s* and *t*, τ_{lim}, are equal to the critical shear stress obtained from the lap shear tests, or 7.4 MPa in the present case. Damage starts to develop at the interface if:

$$\left(\frac{\sigma_n}{\sigma_{\lim}}\right)^2 + \left(\frac{\tau_s}{\tau_{\lim}}\right)^2 + \left(\frac{\tau_t}{\tau_{\lim}}\right)^2 = 1 \quad (2)$$

Next, damage development within the adhesive interface was considered through a strain energy and linear evolution approach. For that purpose the simplified formulation available at item D300, section Conclusion of DNV RP gives an estimative figure. The quasi-static bondline load resistance capacity can be estimated from the strain energy release rate, *G*:

$$G = \frac{2.f^2}{3} \frac{K_{steel}}{K_{lam}.(K_{steel} + K_{lam})},$$

$$K_{steel} = E_{steel}.t_{steel} \quad , \quad K_{lam} = E_{lam}.t_{lam} \quad (3)$$

3. DISCUSSIONS

As the DLS specimens had a −45°/+45° carbon fiber arrangement along the axis, E_{lam} was of about 12GPa. Formula (3) indicated a maximum G of 3600 J/m^2, which proved too high in practice. Experimental results were better fit for a G value of 1600 J/m^2. Figure 3 shows the results of FEM models of DLS tests with long (160 mm) and short (45 mm) overlap lengths, in terms of unit load versus displacement. Also shown are the experimental results for other overlap lengths.

The model behavior resulted somewhat stiffer, but the non-linear nature of interface behavior was adequately modelled. A failure load of about 700 N/mm for long overlaps was indicated, which is close to the experimental average. The reduction in ultimate load for short overlaps was also captured by the model.

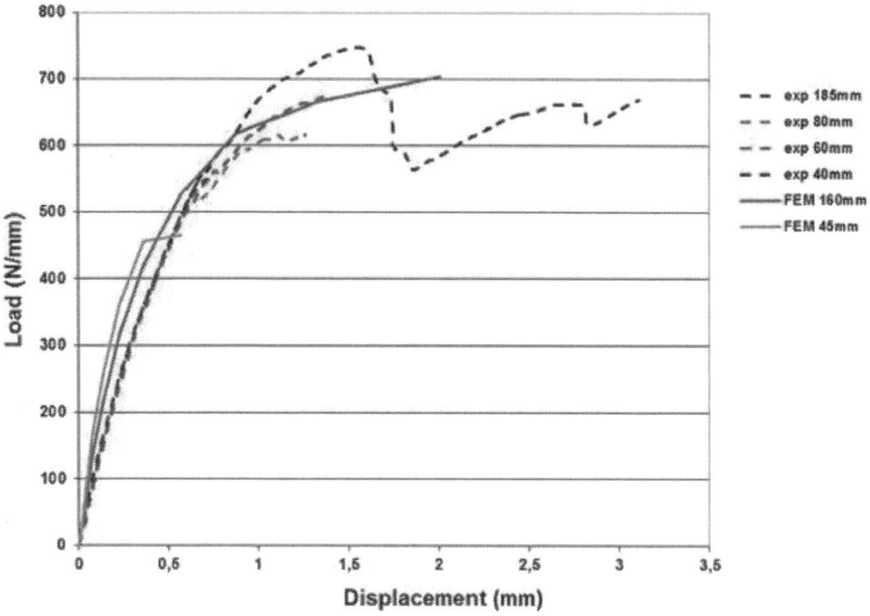

Figure 3. DLS tests, experimental and simulated.

3.1 Experimental verification of repair effect on shear stresses

As the main objective of the repair was to reinforce the hull regarding shear stresses, after the lap shear evaluations three points bending tests were performed on 6 inch I-beams, with and without composite reinforcement. The composite patch was applied to the web of one beam, with a +45°/-45° fiber alignment for optimal shear reinforcement. Eighteen layers were deployed, resulting in 9.2 mm of structural carbon laminate and a 0.6 mm thick adhesive layer. The capacity of the load frame limited the size of the beams. Nevertheless, the analysis concept remains the same no matter how large or small the metallic structure is. Figure 4 shows the reinforced beam sketch.

Both beams were monitored by rectangular strain gage (SG) rosettes applied at the position shown in Figure 4, in order to evaluate the stress field at that point of maximum shear stress. Figure 5 shows both beams at the three points bending experimental setup.

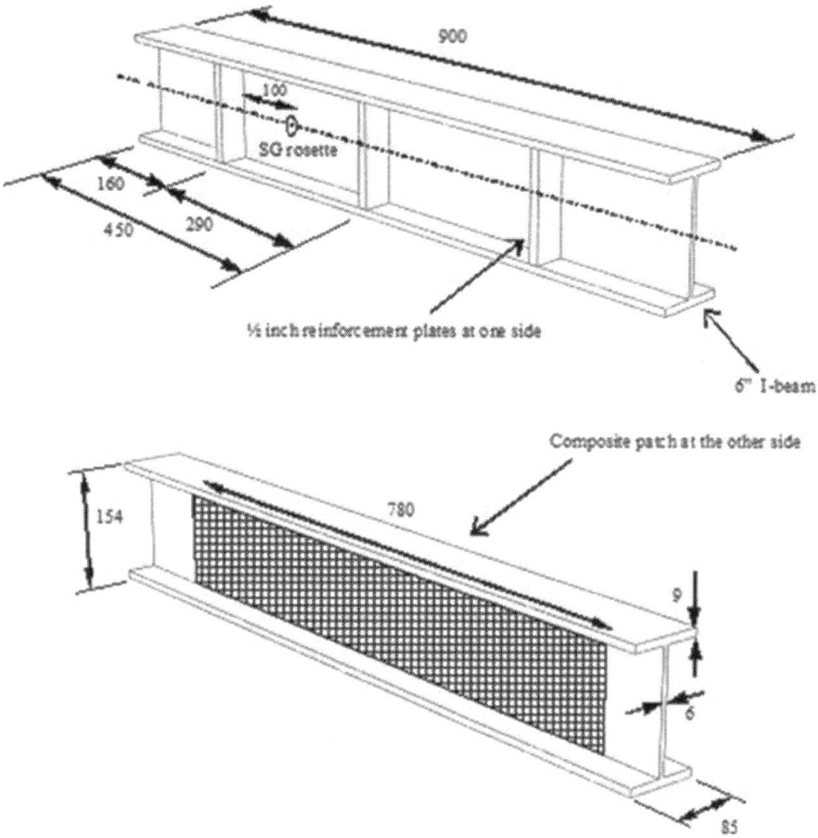

Figure 4. Six inch I-beam with carbon laminate applied to the web.

Figure 5. Experimental setup for three points bending of the beams.

The test results, in terms of loads, displacements and stresses graphs are shown in Figure 6. There was a nonlinear displacement behavior at the beginning of the tests, due to gap closure and geometric accommodation of the beams to the test rig. A plastic load regime started to develop towards the ends of both tests, notably for the unreinforced beam.

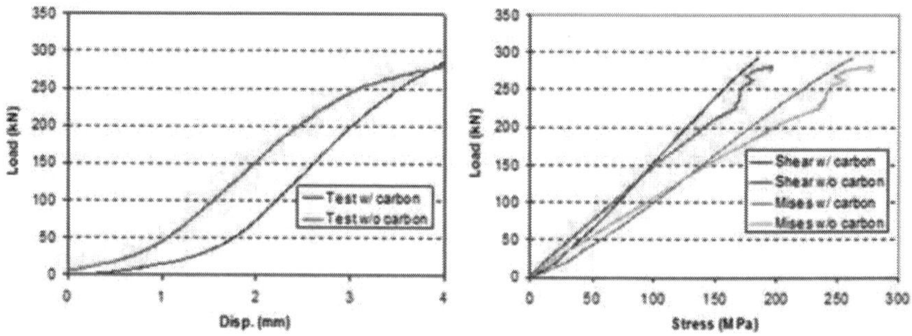

Figure 6. Three points bending tests results.

The SG rosette applied to the unreinforced beam displayed a beginning of plasticity and some erratic behavior above 150kN load level. Nevertheless, an elastic regime was captured between 100kN and 150kN for both tests, so this load range provided the basis for results comparison. Between those two load levels there was an increase in shear stress of 35 MPa for the unreinforced beam and of 27 MPa for the reinforced beam. In conclusion, the 9 mm thick carbon laminate bonded to the web caused a 23% reduction of shear stresses at that instrumented point.

3.2 Defect fatigue propagation curve definition

3.2.1 Determination of strain energy release rates

Consider a defect of typical size a at the adhesive interface. The basic parameter for fatigue analysis of the adhesive layer, associated to defect propagation (disbondment) is the strain energy release rate (SERR), G, already mentioned. It is defined in fracture mechanics as the strain energy dissipated per unit of newly created defect areas. It can be obtained from FEM analysis by computing the difference in total strain energy stored as the component is deformed, divided by the increment in defect area, for the geometries before and after a small defect growth, da :

$$G = \frac{U_a - U_{a+da}}{A_a - A_{a+da}} \qquad (4)$$

For the present DLS test setup, G was evaluated for several different defect sizes, namely 20, 40, 70 and 80 mm, selected to leave an overlap length still greater than the maximum effective, 100 mm. The models simulated a 5% increase in defect area and the SERRs were computed for the defect sizes indicated in formula (4). The models results, relating the unit load f to G are shown in Figure 7.

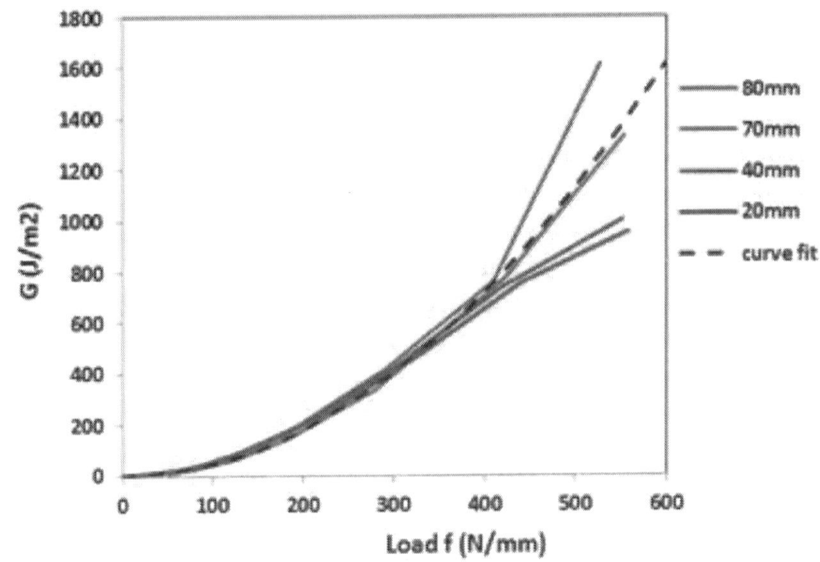

Figure 7. Correlation between load and SERR.

As can be seen, strain energy varies with the square of unit loaf f - in agreement with formula (3) - up to f around 400 N/mm, which is little more than half the quasi-static failure load. This range is also within the approximately linear behavior of the interface, indicated by the load x displacement curve, as shown in Figure 3. Moreover, within this load range *G* is independent of defect size and a parabolic curve fit, shown as a dashed line in Figure 7, gives a conservative estimative of *G* values for defects up to 70 mm in size, even for load levels above 400 N/mm.

3.3 Fatigue tests

Having determined the relation between *f* and *G*, several defect propagation DLS tests were done at a servo-hydraulic test machine, keeping constant the maximum load and a load ratio of 0.1. For the fatigue tests the carbon laminates were not cut, in order to provide sufficient length for defect propagation and still allowing enough room for the maximum effective overlap length of 100 mm. The frequency was 10 Hz and laminate temperatures were controlled in order to avoid over-heating. A white paint was applied to the side of the specimens, as shown in Figure 8, to make defects visible.

In the beginning there were no initial defects but as soon as the tests started there was a rapid defect nucleation at the center of the specimens, followed by stable propagation. As discussed above, for defects up to 70 mm in length, *G* is considered independent of defect size and is obtained from the unit load *f* through the dashed curve of Figure 7. The parameter adopted for fatigue evaluation was the maximum *G* reached in each cyclic test [7]. As the specimens were tested in pairs, each test provided four defect fatigue propagation results. Defect size was the average of two defect tip measurements, from the front and back faces of specimen.

Fatigue test results are displayed at Figure 9, as a log-log diagram of G_{max} *versus* defect propagation rate da/*dN* (Paris law). As indicated at the graph, a threshold of $Log(G_{max}) = 2.5$ was assumed. This corresponds to an extremely low propagation of one nanometer per cycle and means that defects providing G_{max} equal to 317 J/m^2 or less would practically not grow due to fatigue. It is recalled that the fatigue tests were performed under a load ratio of 0.1. The experimental data provided by Alegri et alii [8] indicate that higher load ratios would lead to higher thresholds.

Figure 8. Fatigue test, showing defect propagation.

4. RESULTS

4.1 FEM model of the repaired region

The local model of the damaged hull region was translated to ABAQUS code through an input file obtained from the original model of the platform design database. The area with thickness losses had a refined mesh with 50 mm of shell element size. Metallic plate thickness variation within the model was considered according to a thickness map obtained from hull inspection measurements.

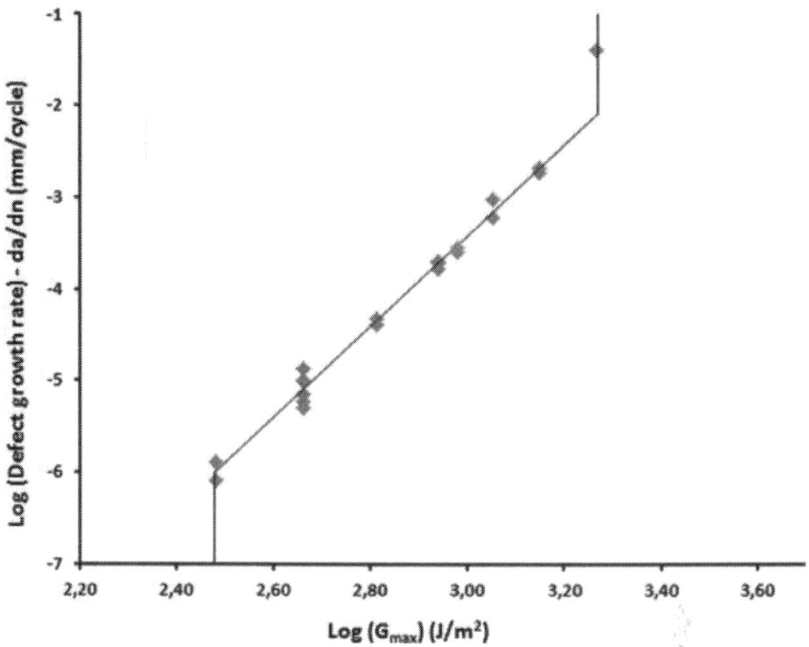

Figure 9. Defect fatigue propagation curve obtained.

The local model had displacements imposed at its boundaries, which were obtained from the global ship analysis. From the several cases studied, load case 4 – a given combination, among several others, of cargo tank levels and extreme storm wave, current and wind loads, including incident directions – that provided the highest shear stresses at the repair location in the hull. Figure 10 shows the local ship model, including the internal reinforcement structure, the damaged hull and the composite repair laminates superimposed to it. The dimensions of the area that needed reinforcement were of about 5×3.5 m.

The basic design drive was to restore the original plate in-plane stiffness, along fiber directions. A +45°/-45° fiber disposition in respect to ship axis was adopted, as explained before, for shear reinforcement. Given E_1 and E_2 moduli of 46 GPa measured for the composite laminate, a 4.6 (210/46) multiplier applies to steel thickness losses to obtain the corresponding carbon thickness. The minimum thickness required by Class at the hull position under analysis is 19.6 mm. For the most affected plating, with 11.5 mm of steel remaining, an added 8.1 mm of steel or 37.3 mm of carbon was thus needed.

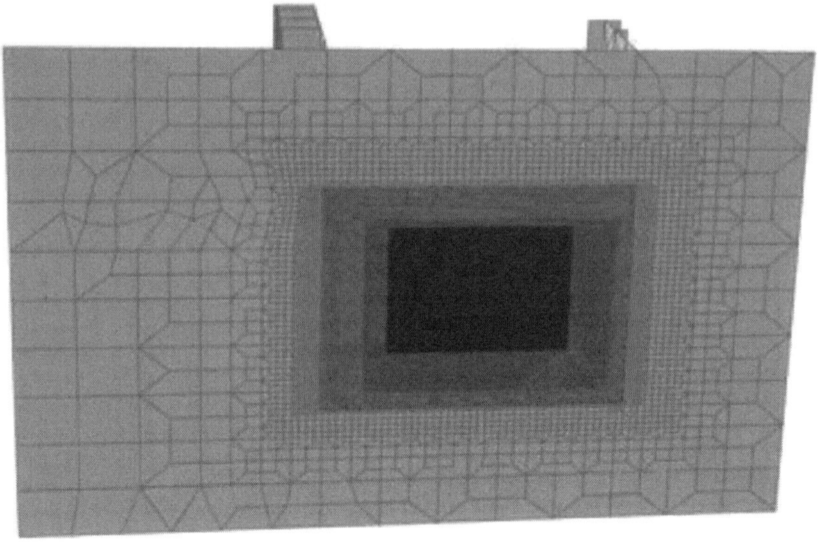

Figure 10. Local model of the repaired region of the hull.

In order to optimize carbon fiber consumption the repair was divided in three parts, as displayed at Figure 10. The characteristics of the 3 laminates are shown at Table 2. The lamination went from the largest layer to the smaller, with size decrements from sheet to sheet to provide thickness tapering at the borders.

Table 2. Characteristics of the repair laminates

#	Length (m)	Height (m)	Number of layers	Carbon thick. (mm)	Equiv. steel thick. (mm)
1	5.5	4.0	30	18.0	3.9
2	4.4	3.1	17	10.2	2.2
3	3.3	2.3	16	9.6	2.1

The total carbon thickness effectively deployed was 37.8 mm. The adhesive interface properties discussed in section FEM simulation of the DLS tests were introduced into the hull repair model, for load case 4. As already explained, the composite repair FEM grid is simply superimposed to the steel one. The plate element size for the composite repair is 25 mm. The results, in terms of Tresca stress invariant, are shown at Figure 11, for the inner laminate surface, in contact with the interface.

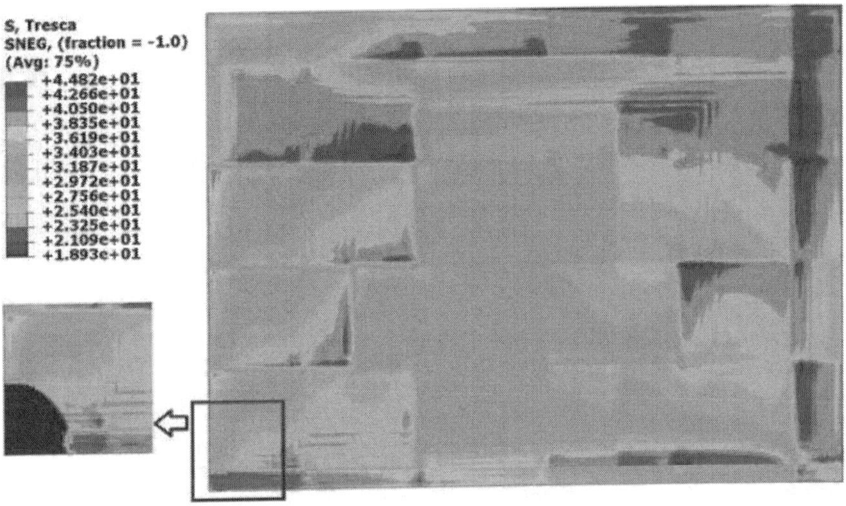

Figure 11. Tresca stress invariants, inner lamina (MPa).

After the repair stress field was obtained for this load case, some circular defects were simulated at the rightmost, lower repair corner, where the Tresca stress invariants were largest, as indicated by the inset at Figure 11. The defects simulated disbondments between steel and composite, starting at the edges. Two defect sizes were simulated: 200 and 500 mm in radius. Then, similarly as it was done for the simulation of DLS tests, the defects were considered to have grown about 10% in area.

The strain energies given by the models both before and after defect growth were obtained and the SERRs were calculated as indicated by expression (4). For the smallest defect it resulted to be 69.7 J/m^2 and for the largest, 67.4 J/m^2. The load case in study is a maximum one, with very few occurrences, but even if it were frequent the defects would not grow in fatigue, as the threshold is 317 J/m^2.

4.2 Structural monitoring results

The strain monitoring system adopted Bragg grating optical strain gages, in order to eliminate zero drifts and electromagnetic interferences that could be captured by the long cable needed to drive the signals from the hull to the local processing and data transfer unit, located at the platform deck. A total number of thirteen delta strain gage rosettes were applied at the external surface of the composite patch repair, together with two dummy sensors for temperature compensation. Strain data was acquired four times a day and sent to the office in Rio de Janeiro through the company intranet.

5. DISCUSSIONS

Data processing indicated that dynamic stresses experienced by the repair along 2013 were well below the maximum values indicated the FEM model for the load case show at Figure 11, as in this case the maximum Tresca stress invariants at positions where strain gages were installed at the composite repair was 36 MPa. The histogram of Figure 12 shows high frequency (periods from 5 to 15 s) Tresca stress ranges at the repair, measured by sensor number 10 along July 2013, when they were the highest.

The histogram above is related to loads caused by short term wave, wind and current loads. As the monitoring system is based in optical strain gages, long term variations due to oil tanks levels, temperature changes, ship weathervane, sea states, storms, etc. were also captured. It is shown at Figure 13, which displays average Tresca stress measured by sensor 10, from March to December 2013.

It can be seen from the graph that long term stress ranges sometimes approached the maximum load case illustrated in Figure 11, but as shown before, it is still well below the fatigue threshold. So, the monitoring results also indicate that eventual defects that exist at the metal/composite interface will not grow due to fatigue.

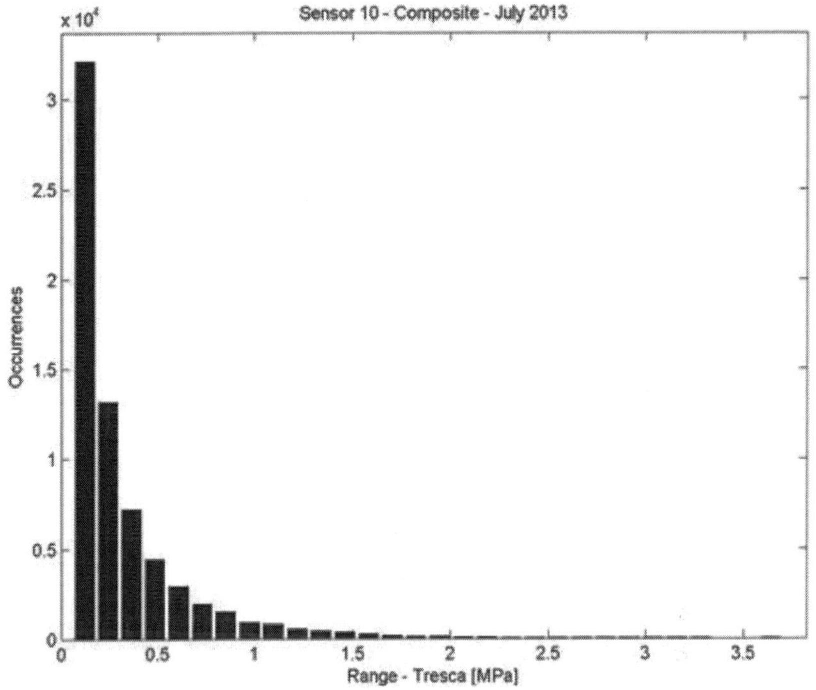

Figure 12. Short term Tresca stress ranges measured.

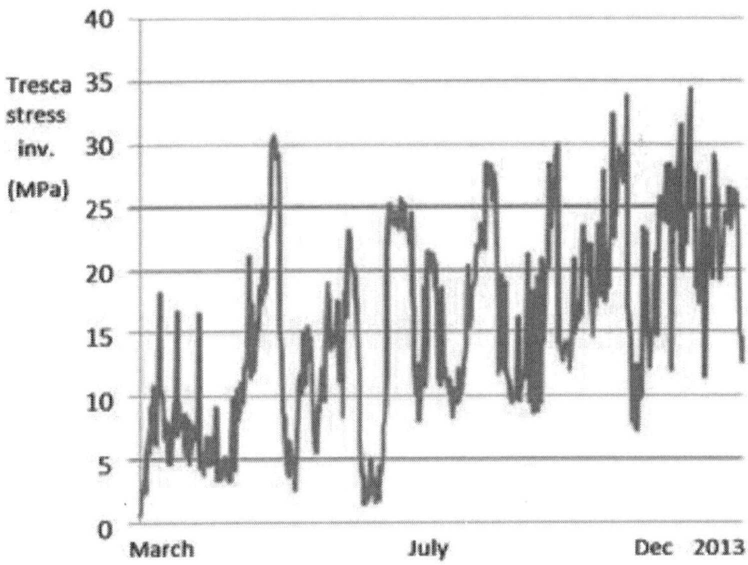

Figure 13. Long term Tresca stress invariant averages.

6. AGING

Part of the original reinforced plate from which the DLS test specimens were cut was reserved during the materials qualification period at the beginning of the project. Afterwards, it was submitted to an accelerated aging program in an environmental chamber, firstly with one week long exposition to salt spray, followed by another week long exposition to UV radiation. This phase lasted for six months.

At the end of that period there was not any visual indication of degradation of the reinforced plate, so only the UV exposition remained, and lasted for eight months more. After that, three extra DLS tests were performed, in order to address any modification of the interface behavior. These specimens had long overlaps, i.e., there were not any cuts in the carbon laminates. These test results are shown in red at Figure 14, superimposed to the original ones, shown in grey color at the graph.

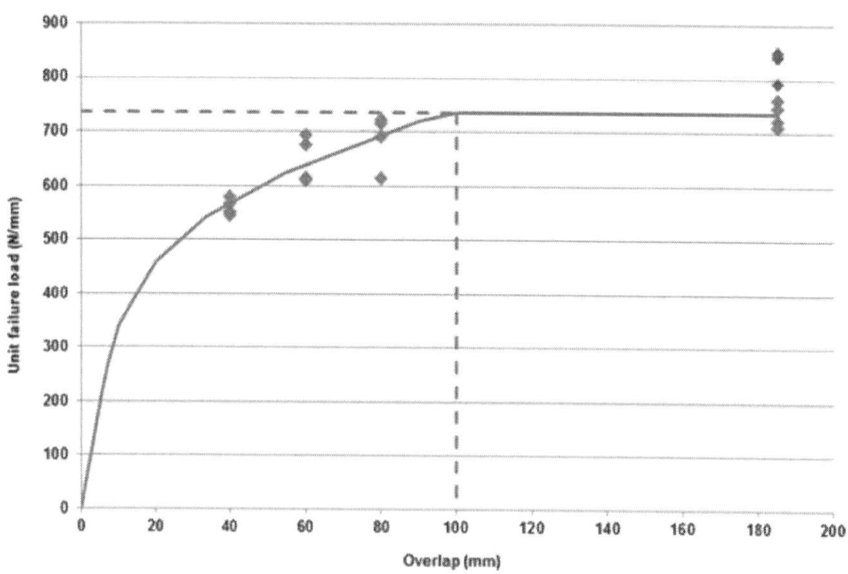

Figure 14. DLS test results after aging (in red).

As can be seen, the aging process caused an improvement in interface properties, as the unit failure load increased from 736 N/mm to an average value of about 830 N/mm, a 12% increase factor. What the aging test has shown is that the repair itself provided an efficient barrier against the environment and protected the interface from any chemical or physical attack. The improvement of unitary failure load with time thus indicates a completion of the adhesive curing process. It can be concluded that the safety factor of the patch repair against instantaneous disbondment is even higher than originally designed, after environmental aging.

7. CONCLUSIONS

A FEM adhesive interface behavior model was established from the mechanical properties of both adhesive and composite materials and DLS test results. Defect propagation tests provided a suitable defect growth fatigue curve. Both interface model and fatigue data were utilized in the FEM modelling of the repair executed, and together with the strain monitoring data acquired, they led to the conclusion that the repair will not fail due to fatigue propagation of eventual defects existing at the adhesive interface. Furthermore, environmental aging was beneficial as it caused a 12% increase in the critical shear stress of the interface.

COMPETING INTERESTS

The authors declare that they have no competing interests.

AUTHORS' CONTRIBUTIONS

LM coordinated the repair operation, performed the characterization of adhesive interface properties, the experimental verification of repair effect on shear stresses and executed the FEM modeling. LL was in charge of the fatigue tests. SM provided the instrumentation of the repair with optical strain gages and operated the monitoring system. All authors read and approved the final manuscript.

REFERENCES

1. Grabovac I, Whittaker D: "Application of bonded composites in the repair of ships structures – a 15-year service experience". Composites Part A 2009, 40: 1381–98.

2. Det Norske Veritas:"Technical Report - Project Recommended Practice for Composite Patch Repair of FPSO Structures", 1st edition, Publisher. 2006.

3. Meniconi LCM, Porciúncula IN, McGeorge D, Pedersen A: Structural Repair at a Production Platform by Means of a Composite Material Parch". Offshore Technology Conference, Houston; 2010.

4. Echtermeyer A, McGeorge D, Grave J, Weitzenbock J: "Bonded patch repairs for metallic structures – a new recommended practice". J Reinforced Plastics Composites 2014, 33(6):579–85.

5. Dassault Systèmes Simulia Corp:ABAQUS version 6.13 Users Guide, Publisher. 2013.

6. American Society for Testing and materials:"ASTM D 3528 Standard Test Method for Strength Properties of Double Lap Shear Adhesive Joints by Tension Loading", Publisher, reapproved. 2008.

7. McGeorge D: "Inelastic fracture of adhesively bonded overlap joints". Eng Fracture Mech 2010, 77: 1–21.

8. Allegri G, Jones MI, Wisnom MR, Hallett SR: "A new semi-empirical model for stress ratio effect on mode II fatigue delamination growth".Composites Part A 2011, 42: 733–40.

CHAPTER 2

Numerical investigations of ship–ice interaction and maneuvering performance in level ice

Quan Zhou, Heather Peng *, Wei Qiu

Faculty of Engineering and Applied Science,
Memorial University of Newfoundland, St. John's, Canada

ABSTRACT

This paper presents a solution to the ship–ice interaction problem solved in the time domain by a combined method involving numerical simulations and semi-empirical formula. The breaking process of an intact level ice by an advancing ice breaker has been investigated by a 2D numerical method. An interaction detection technique has been introduced. In order to compute the ice load, a pressure-area relationship for the ice was applied, and a flexural ice plate model was developed. A 3-degree-of-freedom (3DOF) model was adopted to simulate ship maneuvering in level ice. The proposed numerical method has been validated by using the experimental results of an 1:20

scaled CCG R-Class icebreaker model and the full-scale turning circle results.

KEYWORDS

Ship–ice interaction; Ice breaking; Contact detection; Flexural ice plate; Ice loads in 3DOF; PMM tests

1. INTRODUCTION

The growing interests in oil and gas exploration in Arctic and sub-Arctic regions and in the transportation through the Northern Sea Route have led to extensive research on better understanding of ship–ice interaction and vessels' maneuvering performance in ice covered water. For a ship advancing and turning in level ice, the assessment of ship maneuvering performance needs to be addressed. For a ship advancing in ice, the primary interest is on the prediction of the ice resistance. For maneuvering studies, the transverse force and turning moment are as important as the resistance when the ship is in turning.

It is challenging to predict the resistance that a ship is encountered in the intact ice field. One of the common assumptions, the principle of superposition to the total resistance is widely accepted by many researchers when the global ice load models were developed (for example, Enkvist, 1972, Kashteljan et al., 1969, Keinonen, 1996, Lewis and Edwards, 1970, Lindqvist, 1989, Milano, 1972, Riska et al., 1997 and Spencer, 1992). In the past decades, efforts have been made to improve these models and implement them in numerical methods. For instance, Wang (2001) adopted the conceptual framework of the nested hierarchy of discrete events, proposed by Daley, 1991 and Daley, 1992 andDaley et al. (1998), and simplified the method by considering three continuum processes of crushing, bending, and rubble formation. A geometric grid method, which requires the discretization of the entire computational domain, was proposed to simulate continuous contacts between the structure and the level ice. The mechanics of ice crushing-bending failure was applied and the ice loads were numerically computed. This approach has been adopted and modified by many other researchers, such as Nguyen et al. (2009),Sawamura et al.

(2009), Su et al. (2010) and Zhou and Peng (2014). Tan et al. (2013)proposed a 6 degree-of-freedom (DOF) ship–ice interaction model by discretizing a belt area around the ship waterline. Their work provided a possibility to include ship heave, roll and pitch motions during the icebreaking process. A further study on the effect of dynamic bending of level ice during ship–ice interaction was carried out with this 6 DOF model byTan et al. (2014).

Although the fundamental mechanism of icebreaking is not fully understood, many semi-empirical methods were proposed to estimate the resistance of a vessel in ice. However, limited effort has been made on the estimation of the yaw moment for a ship maneuvering in level ice. Lau et al. (2004) proposed a method to estimate the total yaw moment that is analogous to that of ice resistance in the work of Spencer (1992). The total yaw moment was divided into hydrodynamic, breaking, submergence, and ice clearing components. The formulas for the terms associated with breaking and submergence were presented. In their work, the ice-induced forces were considered as three concentrated loads among which two were acting at the bow and the other was on the midship section. The yaw moment was obtained by multiplying those forces and the corresponding lever arm lengths. Martio (2007) developed a numerical program to simulate the vessel's maneuvering performance in uniform level ice based on Lindqvist's ice resistance model (Lindqvist, 1989). The major contribution was to extend the analytical formulas to 3 DOF. The effects of bending and submergence forces on sway and yaw motions were taken into account. A relationship between the resistance and the yaw moment, as well as the transverse force, was developed. Only crushing was calculated numerically, and the other two terms were obtained by using analytical formulas. Nguyen et al. (2009) and Su et al. (2010) developed 3DOF ship–ice interaction models. In their work, it was assumed that only the term associated with breaking had effect on sway and yaw motions. The remaining components of ice loads were considered as resistance that only affected the surge motion.

When a ship is maneuvred in level ice, the ship–ice interaction can be described as a continuous process. As the ship moves into the ice sheet, local shear fracture, crushing failure, and vertical deflection occur at the edge. These ice failure modes manifest themselves as a breaking force on the hull. When the stress somewhere inside the ice sheet exceeds its limit, a bending failure will occur which leads to the breaking of ice floes. The broken ice pieces are further accelerated and rotated by the ship until they are parallel to the hull. Subsequently, the broken pieces slide along the hull until they lose the contact with the ship. During this process, two force components are assumed: the buoyancy caused by the density difference between the ice and the water, and the clearing force that is attributed to the accelerating and rotating ice floes as well as the frictional force during ice sliding. The total ice resistance therefore consists of these three components discussed above. In this paper, the breaking process was simulated by a 2D numerical method involving the discretization of ice edges and ship's waterline. The contact loads were calculated, compared with the bending capacity of the ice sheet, and were used to determine the possibility of ice breaking. Equations of ship motions were then solved, and the ice edges were updated. The clearing and buoyancy forces were calculated using published empirical formulas. The effects on sway and yaw motions were taken into account in this paper. A flexural ice plate model was also developed to modify ice loads. Experimental results of the PMM tests of an 1:20 scale R-Class icebreaker ship model were used to validate the numerical method.

2. MODELING OF SHIP-ICE INTERACTION

2.1. Coordinate systems and mathematical formulations

As shown in Fig. 1, three coordinate systems are employed in the computations. The earth-fixed coordinate system is denoted as *Oe–XeYeZe*. The position and heading of a ship is described in this coordinate system. The ship-fixed coordinate system, *o–xyz*, is with *ox*-axis pointing to the bow, *oy*-axis pointing starboard and *oz*-axis positive downward. The origin, *o*, is located at the interaction point of the longitudinal central plane, the waterplane and the midship section. Equations of ship motions are solved in the ship-fixed coordinate system; The ice-fixed coordinate system, *oi–xiyizi*, is on the ice plate and parallel to *Oe–XeYeZe*. The coordinates of the discretized points on ice edges are defined in *oi–xiyizi*.

The equations of motion for maneuvering, including surge, sway and yaw, are given below:

$$\begin{cases} \dot{x} = \cos(\psi)u - \sin(\psi)v \\ \dot{y} = \sin(\psi)u + \cos(\psi)v \\ \dot{\psi} = r \end{cases} \tag{1}$$

$$\begin{cases} m\dot{u} - mvr - mx_G r^2 = X_H + X_P + X_R + X_{ice} \\ m\dot{v} + mx_G\dot{r} + mur = Y_H + Y_R + Y_{ice} \\ I_z\dot{r} + mx_G\dot{v} + mx_G ur = N_H + N_R + N_{ice} \end{cases} \tag{2}$$

where m and Iz are the mass of the ship and the mass moment of inertia, respectively;xG is the center of gravity defined in the ship-fixed coordinate system; X, Y and Ndenote the forces and moment with respect to ox, oy and oz-axis, respectively; x and $y$are the coordinates; ψ is the orientation; u, v and r denote the corresponding velocities in the ship-fixed coordinate system; the overhead dot represents the time derivative; variables with the subscripts, H, P, R and ice, represent forces and moment due to hydrodynamic force, propeller force, rudder force and ice load, respectively.

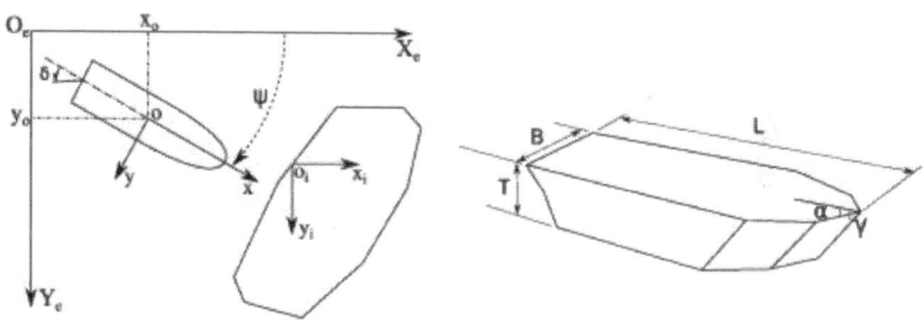

Figure 1. Three coordinate systems.

2.2. Ice-induced forces

The following assumptions were made in order to simplify the hull–ice interaction problem:

- The intact ice sheet is semi-infinite with uniform thickness. It is stationary with respect to *Oe–XeYeZe*;
- The hull–ice interaction is a continuous process which involves repeating cycles of contacting, crushing and bending and breaking;
- Superposition principle is applied to the total ice induced resistance;
- Vertical displacement of the ship is neglected;
- Contacting surfaces remain flat during crushing;
- The shape of the broken ice floe is considered circular for computing in the bending failure.

The total resistance due to ice can be written as:

$$\begin{cases} X_{ice} = X_{br} + X_{cl} + X_{buoy} \\ Y_{ice} = Y_{br} + Y_{cl} + Y_{buoy} \\ N_{ice} = N_{br} + N_{cl} + N_{buoy} \end{cases} \tag{3}$$

The breaking force, denoted as **Fbr** = [*Xbr*, *Ybr*, *Nbr*]*T*, exists when the ice plate is crushed and bent downward until the breaking occurs. The dynamic clearing force, **Fcl**, acts on the hull as the broken ice floes rotates, accelerates, submerges and slides along the ship. The static force due to the buoyancy of ice is denoted as **Fbuoy**. The open-water resistance is considered as hydrodynamic force and is not part of ice loads. Each term in the equation above has effect on surge, sway and yaw motions. The breaking term is calculated numerically based on the detected contacts at each time step. The resistances due to clearing and buoyancy (*Xcl* and *Xbuoy*) are calculated using published empirical formulas. The relationship between *X* and *N* in the clearing and buoyancy terms is derived in the following sections.

2.2.1. Ice breaking force

In the work of Zhou and Peng (2014), a numerical method was proposed to study ship maneuvering problems in level ice. A 2D interaction detection technique was introduced, and the calculation of ice breaking force was provided with details. This study is a continuation of the previous work and the same detection technique was applied. A brief summary of the method and its numerical implementation are presented below.

The waterline of the ship is discretized as a closed polygon while an ice edge is discretized as broken lines (Fig. 2). Note that this method is different from those proposed by Nguyen et al. (2009), Sawamura et al. (2009) and Wang (2001). Instead of discretizing the entire ice sheet, the present method only focuses on the ice edge where the ship–ice interaction may occur, which leads to improved computational efficiency.

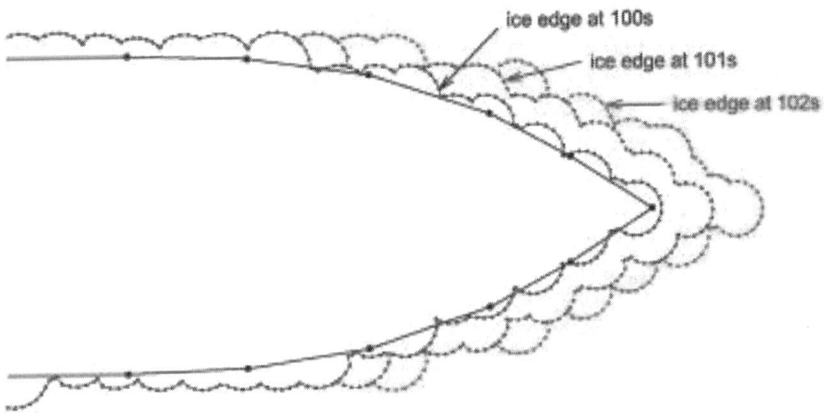

Figure 2. Discretization of ship waterline and ice edges.

Four contact detection scenarios as shown in Fig. 3 were considered, including (1) no ice node is inside the ship and no ship node is inside the ice, (2) ice nodes are inside the ship and ship nodes are inside the ice, (3) ice nodes are inside the ship but no ship node is inside the ice, (4) no ice node is inside the ship but ship nodes are inside the ice. They were distinguished by the Point-in-Polygon method which determines the relationship between a point and a closed polygon. For example, if an arbitrary ice node is inside the polygon of the ship waterline, Scenario 2 or 3 will occur. Then all the nodes on the ship waterline need to be checked. If all ship nodes are inside the ice sheet, Scenario 2 needs to be considered; otherwise, Scenario 3 occurs. Scenario 4 could be identified similarly if no ice node is inside the ship waterline and the ship nodes are inside the ice sheet. As the high fidelity and the computational efficiency need to be satisfied, it is crucial to maintain the accuracy for Scenario 4 while using relatively coarse mesh to discretize the ice edge. Once a contact scenario was found, interpolations were applied to identify the intersections and apexes of the ideal ice wedge, which is illustrated by the blue dash lines in Fig. 3. The open angle of the ice wedge is denoted as θ. The breaking force was calculated based on the interaction area and the ideal ice wedge. If the force exceeds the bending limit of the ice sheet, a circumferential crack will occur and a floe will be broken off the ice sheet. The newly formed ice edge will be discretized into nodes and used to replace the nodes that are enclosed by the arc. This process is known as the ice edge update. The flow chart for the numerical implementation of the algorithm is given in Fig. 4.

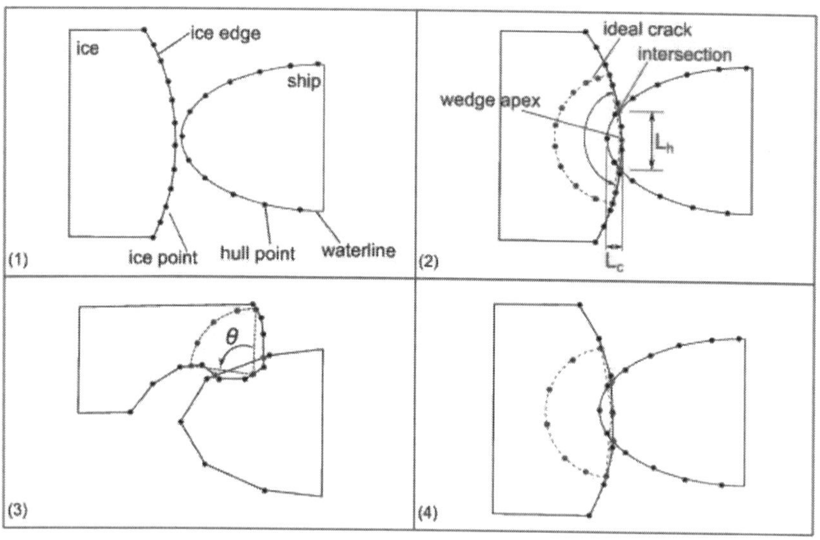

Figure 3. Contact detection scenarios.

As the hull contacts with ice edge, ice crushing starts at the contact point. It continues until the ice plate is broken by bending moment. More than one contact zone could exist at the same time. In each contact zone there will be a crushing force acting on the hull. To determine the magnitude of the crushing force in each contact zone, the following equation is adopted:

$$F_{cr} = p_{ave}A_{cr} \tag{4}$$

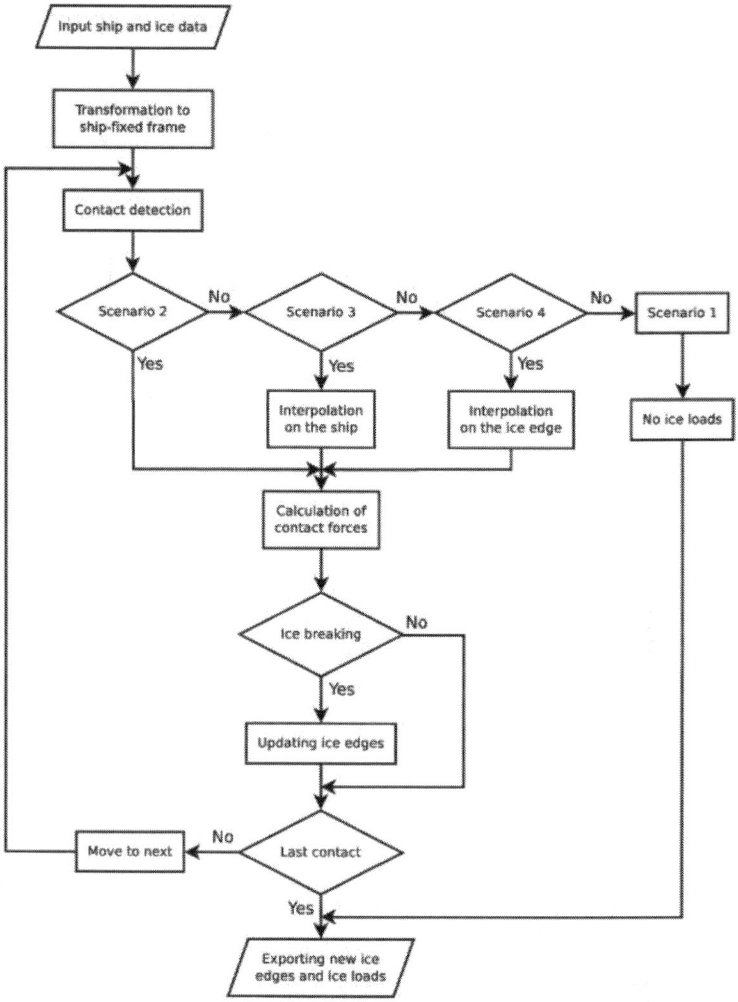

Figure 4. Flow chart.

where *Fcr* is the crushing force, which is normal to the contact surface and pointing inward to the ship; *pave* is the average pressure on the contact surface; *Acr* is the area of the contact surface that is obtained numerically at each time step, which is calculated by

$$A_{cr} = \begin{cases} \frac{1}{2} L_h \frac{L_c}{\sin(\beta')} & \text{if } L_c \le h_i \cdot \tan(\beta') \\ \frac{1}{2} \left[L_h + L_h \frac{L_c - h_i / \tan(\beta')}{L_c} \right] \frac{h_i}{\cos(\beta')} & \text{if } L_c \ge h_i \cdot \tan(\beta') \end{cases} \quad (5)$$

where Lh is the characteristic width of the overlap area; Lc is the indentation; hi is the ice thickness; Lh and Lc are numerically determined at each contact point when the overlap area has been identified; and β' is the normal frame angle of the ship hull that is determined by

$$\begin{cases} \tan(\beta') = \tan(\beta) \cdot \cos(\alpha) \\ \tan(\beta) = \tan(\alpha)/\tan(\gamma) \end{cases} \tag{6}$$

where α and γ are the waterline angle and the buttock angle, respectively, as shown inFig. 1; and β' is the normal section angle.

In the previous work, for example, Lau et al. (2004), Martio (2007), Nguyen et al. (2009),Sawamura et al. (2009), Su et al. (2010) and Zhou and Peng (2014), a constant contact pressure (crushing strength, σcr) was used when calculating the crushing force. However, the analysis of full-scale data shows that the contact pressure varies as the nominal contact area changes. This phenomenon can be expressed by the pressure-area relationship (also known as the *P-A* curve) and a form of power relation as below is widely accepted:

$$p_{ave} = p_0 A_{cr}^{ex} \tag{7}$$

where p_0 is assumed equal to the crushing strength and here ex is chosen as − 0.5.

Based on the work of Zhou and Peng (2014), the three components of ice breaking force are calculated by:

$$X_{br} = \left(\sin(\beta')\tan(\alpha) + \mu\sqrt{1 + \tan^2(\alpha)\cos^2(\beta')} \right) \cdot F_{cr} \tag{8}$$

$$Y_{br} = \left(\sin(\beta') - \mu \tan(\alpha) \frac{\cos(\alpha) - \cos^2(\beta')}{\sqrt{\cos^2(\alpha) + \sin^2(\alpha)\cos^2(\beta')}} \right) \cdot F_{cr} \tag{9}$$

$$Z_{br} = \left(\cos(\beta') - \mu \frac{\sin(\alpha)\sin(\beta')\cos(\beta')}{\sqrt{\cos^2(\alpha) + \sin^2(\alpha)\cos^2(\beta')}} \right) \cdot F_{cr} \tag{10}$$

where *Xbr* and *Ybr* are the force components along *x*-axis and *y*-axis, respectively; *Zbr*is the vertical component that is further compared to the bearing capacity of the ice plate to determine if the breaking process would occur.

The yaw moment is obtained from:

$$N_{br} = -X_{br} \cdot y + Y_{br} \cdot x \tag{11}$$

where *x* and *y* are the coordinates of the contact location with respect to *o-xyz*.

It should be pointed out that Eqs. (8), (9), (10) and (11) give the breaking force at one contact location. The global breaking force can be calculated by summing the forces in different locations.

2.2.2. Crushing force adjustment

Valanto (1989) reported that a rapid flexural failure was observed in the experiment when the flexural strength ratio was large ($E/\sigma f = 6400$), while considerably more time was required for flexural failure in the low ratio case ($E/\sigma f = 1400$). It was also pointed out that the difference in two cases was important since it would significantly affect the average ice resistance. In the work of Nguyen et al. (2009), Su et al. (2010) and Wang (2001), rigid ice plate models were assumed and the vertical deflection by bending was not considered. Their models were thus unable to simulate what was observed in the experiment byValanto (1989). In this study, the ice plate flexural deflection was taken into consideration, and an adjustment factor of contact surface was derived.

The actual ship–ice interaction is a dynamic process as both the ship and the ice plate have vertical motions. However, the process was simplified in this work by assuming that the ship does not move vertically. As the ship moves forward, the elastic ice plate will be bent. Since the ice deflection is small compared to the characteristic length of the ice, a parallel downward movement of the ice plate was further assumed and there is no bending involved (see Fig. 6). The contact force, *Fcr*, will lead to ice crushing and bending simultaneously. The water support was considered as an elastic foundation.

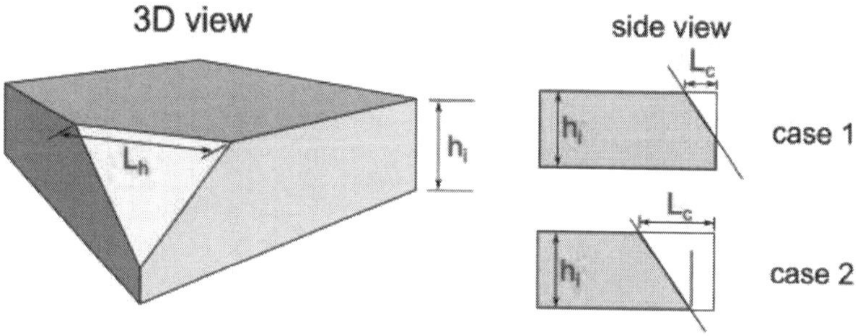

Figure 5. Ideal ice wedge and contact surface.

As shown in Fig. 6, from the initial contact to the present time step, the ship penetrated into ice by δs, which was determined numerically, causing a vertical elastic deflection, δe, and a vertical crushing height δc, which satisfy the following relationship:

$$\delta_e + \delta_c = \delta_s / \tan(\beta') \tag{12}$$

Both δe and δc are unknown. Both are induced by the same force, the vertical component of the contact force, *Fcr*.

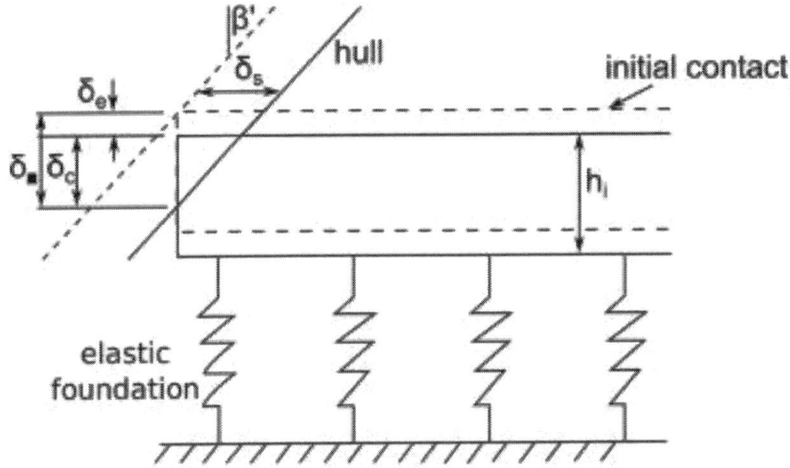

Figure 6. Flexural deflection of ice plate.

The relation between δe and *Fcr* is derived by considering the response of a homogeneous and isotropic elastic ice plate with an open angle of θ, which rests on a liquid and is subjected to a static vertical concentrated load *P* on the apex. Kerr (1976)proposed the following formula to calculate the deflection at the apex of a semi-infinite plate with an arbitrary open angle:

$$w_0 = \frac{1}{2}\left(\frac{\pi}{\theta}\right)^2 \frac{P}{\sqrt{\gamma D}} \tag{13}$$

where *P* is the concentrated load; $\gamma = \rho w g$ is the weight per unit volume of the liquid;$D = Ehi^3/(12(1-v^2))$ is the flexural rigidity of ice plate; *E* is the ice Young's modulus; *v*is Poisson's ratio. Substituting $P = Fcr{,}v$ and $w_0 = \delta e$ into Eq. (13), the following relationship can be achieved

$$F_{cr,v} = \frac{2\theta^2}{\pi^2}\sqrt{\rho_w g D} \cdot \delta_e \tag{14}$$

The relationship between Fcr,v and δc is implicitly expressed by Eqs. (4), (5) and (10).

The binary search method was applied to obtain δc. Once it is obtained, the contact areaAcr in Eq. (4) can be modified by multiplying $(\delta c/\delta v)^2$, because the flexural deflection of ice plate will reduce the contact area. Therefore, the adjusted crushing force is:

$$F^*_{cr} = p_{ave} \cdot A_{cr} \cdot \left(\frac{\delta_c}{\delta_v}\right)^2 \tag{15}$$

The effect of the flexural deflection of the ice sheet can be considered by replacing $Fcr$with Fcr^{*} in Eqs. (8), (9) and (10).

2.2.3. Bending failure criterion

The vertical force component increases as the ship penetrates into the ice plate. When it exceeds the bearing capacity of the ice sheet, the bending failure would occur and a circular ice floe would be broken off the plate. The bearing capacity is calculated by Kashtelyan's method (Kerr, 1976).

$$P_{flex} = C_f\left(\frac{\theta}{\pi}\right)^2 \sigma_f h_i^2 \tag{16}$$

where θ is the open angle of the ideal wedge; σf is the flexural strength of ice plate; hi is the ice thickness; and Cf is an empirical coefficient. It should be noted that Eq. (16) is derived from a quasi-static condition. It does not consider the dynamic effect due to varying ship velocities.

Kashtelyan suggested a small value (around 1.0) for *Cf*, while Nguyen et al. (2009) used a value of 4.5 with no explanation. In the work of Su et al. (2010), a study of *Cf* was carried out and a value of 3.1 was finally adopted. The value of *Cf* is adopted as 2.2 in this work and detailed discussions can be found in Section 3.2.

After the vertical force, *Zbr*, and the bearing capacity of ice, *Pflex*, are determined, the next step is to investigate whether the ice wedge will be broken from the ice sheet. If the vertical force exceeds the capacity, i.e., $Zbr \geq Pflex$, the bending failure will occur and a circular ice floe will be broken off the ice plate; otherwise the ice edge will only be crushed. The radius of the broken ice floe is determined by using the formula in the work of Wang (2001):

$$R = C_l \cdot l \cdot (1 + C_v v_{n,2}) \tag{17}$$

where *R* is the radius of circumferential crack; vn_2 is the velocity component normal to the contact surface; *Cl* is the parameter that defines the broken ice size to its characteristic length; *Cv* is a tunable coefficients, and both are discussed in Section 3.2; and *l* is the characteristic length of ice plate given by:

$$l = \left(\frac{E h_i^3}{12(1-\nu^2)\rho_w g} \right)^{1/4} \tag{18}$$

where *E* is Young's modulus of ice; *v* is Poisson's ratio; and *ρw* is the density of sea water.

2.2.4. Buoyancy and clearing forces

The ice resistances due to buoyancy and ice clearing which includes ice rotating and sliding are obtained using the regression model proposed by Spencer and Jones (2001):

$$X_{cl} = 2.03 F_h^{-0.971} \rho_i B h_i u^2 \tag{19}$$

$$X_b = 2.67 \Delta\rho g h_i B T \tag{20}$$

where $F_h = u/\sqrt{g h_i}$ is the Froude number of the ice; u is the advancing speed; B and T are the ship breadth and draft, respectively; ρi is the density of ice; $\Delta\rho$ is the density difference between ice and sea water.

To calculate the y-component of buoyancy and clearing forces, it is assumed that the ice forces are acting on the bow part of the ship and the midship part. It is also assumed the ship is symmetric about the longitudinal centreplane and the ice force acts on the ship symmetrically when the ship moves straight. It is further considered that the bow is formed of two flat planes. Therefore, the directions of buoyancy and clearing forces should be the same as that of the breaking force, i.e.:

$$Y_{i,bow} = \frac{X_i}{2} \cdot \frac{Y_{br}}{X_{br}} = \frac{\sin(\beta') - \mu \tan(\alpha) \dfrac{\cos(\alpha) - \cos^2(\beta')}{\sqrt{\cos^2(\alpha) + \sin^2(\alpha)\cos^2(\beta')}}}{\sin(\beta')\tan(\alpha) + \mu\sqrt{1 + \tan^2(\alpha)\cos^2(\beta')}} \cdot \frac{X_i}{2} \tag{21}$$

where the subscript i is either "cl" or "b", representing the clearing or buoyancy force, respectively. Note that the force is divided by 2 as the force is assumed to act on either side of the bow.

When a vessel starts turning, the forces on the port and starboard sides are not symmetrical anymore. A net transverse bow force exists. The net force results from the yaw rate of the vessel, i.e., the vessel will encounter larger transverse force if it turns rapidly.

$$\Delta Y_{i,bow} = \frac{u \sin(\theta_1)}{\tan(\alpha)} Y_{i,bow} \tag{22}$$

Where

$$\theta_1 = \text{atan2}\left(r \frac{(L_{WL} - L_{bow})}{2}, u\right) \tag{23}$$

where *LWL* is the ship length at the design waterline, and *Lbow* is the length of the bow as shown in Fig. 7.

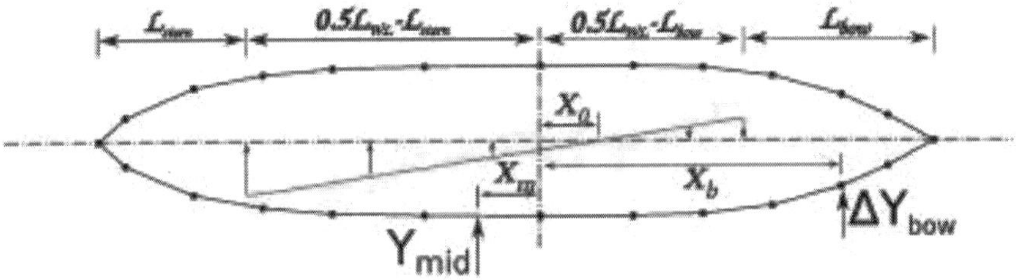

Figure 7. Arms of rotation.

The force acting on the parallel midship section is determined by assuming the load is in the opposite direction of the velocity:

$$Y_{i,mid} = C_{mid} \cdot X_i \cdot \frac{v}{u} \tag{24}$$

where u and v are surge and sway velocities, respectively; and $Cmid$ is equal to 2.0.

After the transverse components of buoyancy and clearing forces, $\Delta Yi{,}bow$ and $Yi{,}mid$, are obtained, the yaw moment can be obtained by multiplying the transverse force by the corresponding lever arm:

$$N_i = \Delta Y_{i,bow} \cdot x_b + Y_{i,mid} \cdot x_m \quad (25)$$

where xm is calculated by

$$x_m = \frac{\frac{1}{3}L_1^2(L_1 r + v) - \frac{1}{3}L_2^2(L_2 r + v)}{\frac{1}{2}L_1(L_1 r + v) + \frac{1}{2}L_2(L_2 r + v)} \quad (26)$$

Where

$$L_1 = \frac{1}{2}L_{WL} - L_{stern} + x_0$$
$$L_2 = \frac{1}{2}L_{WL} - L_{bow} - x_0$$

and $Lstern$ is the length of the stern and $x_0 = -v/r$.

It is assumed that $\Delta Yi{,}bow$ acts on the middle of the bow, i.e.,

$$x_b = \frac{1}{2}L_{WL} - \frac{1}{2}L_{bow} \quad (27)$$

At each time step, the contact will be first detected by the numerical method. If a contact is found, the pressure-area relationship will be applied to determine the pressure. After that, the crushing force is calculated by Eq. (4). The breaking force can be obtained by using Eqs. (8), (9), (10) and (11) and further modified by considering the effect of flexural deflection of the ice plate. Finally, the buoyancy and clearing forces are determined by using Eqs. (19), (20), (22), (24) and (25).

2.3. Hydrodynamic forces

The hydrodynamic forces are determined as follows according to the work of Hirano (1981):

$$X_H = -X_{\dot{u}}\,\dot{\boldsymbol{u}} + X(u) + (m_y + Xvr)vr \tag{28}$$

$$Y_H = -Y_{\dot{v}}\,\dot{\boldsymbol{v}} - Y_{\dot{r}}\,\dot{\boldsymbol{r}} - m_x ur + Y_{H0} \tag{29}$$

$$N_H = -N_{\dot{r}}\,\dot{\boldsymbol{r}} - N_{\dot{v}}\,\dot{\boldsymbol{v}} + N_{H0} + Y_{H0} \cdot x_G \tag{30}$$

Where

$$Y_{H0} = Y_v v + Y_r r + Y_{v|v|} v|v| + Y_{r|r|} r|r| + Y_{vr} vr \tag{31}$$

$$N_{H0} = N_v v + N_r r + N_{vvr} v^2 r + N_{vrr} vr^2 + N_{r|r|} r|r| \tag{32}$$

the constant coefficients, referred to hydrodynamic derivatives, are

$$X_{\dot{u}} = \frac{\partial X}{\partial \dot{u}}, \quad Y_v = \frac{\partial Y}{\partial v}, \quad N_{vvr} = \frac{\partial N}{\partial v^2 \partial r}, \quad \cdots$$

$X(u)$ represents the resistance due to water and is calculated by using the prediction method proposed by Holtrop and Mennen (1982). In this paper, the hydrodynamic derivatives were obtained from the PMM tests with a 1:20 scale ship model.

2.4. Operational forces

The operational forces consist of two parts: the thrust from the main propeller and the lift and drag forces due to the rudder deflection. The model used by Gong (1993) was adopted in this work.

$$X_P = (1-t_P)\rho_w n^2 D_P^2 K_T(J_P) \tag{33}$$

$$X_R = -(1-t_R) F_N \sin(\delta) \tag{34}$$

$$Y_R = -(1 + \alpha_H) F_N \cos(\delta) \tag{35}$$

$$N_R = -(1 + \alpha_H) x_R F_N \cos(\delta) \tag{36}$$

where t_P and t_R are the wake fractions due to propeller and rudder, respectively; n is the propeller revolution per second, (rps); D_P is the propeller diameter; K_T is the thrust coefficient; $J_P = u/(nD_P)$ is the propeller advance ratio; α_H is the coefficient of hydrodynamic force on ship hull induced by the rudder action that can be estimated from model test results; x_R is the longitudinal location of the rudder; δ is the rudder deflection angle; and F_N is the rudder normal force determined by:

$$F_N = \frac{1}{2}\rho_w \frac{6.13\Lambda}{\Lambda + 2.25} A_R U_R^2 \sin(\alpha_R) \tag{37}$$

where Λ is the aspect ratio of the rudder; A_R is the projected rudder area; U_R is the effective rudder inflow velocity; α_R is the effective rudder inflow angle. Different models are available for the calculation of the effective rudder inflow velocity and angle (Gong, 1993, Hirano, 1981 and Kobayashi et al., 2003). The one proposed by Hirano (1981)was employed in this paper.

2.5. Numerical implementation

At each time step, the external forces on the right hand side of Eq. (2) are obtained by the numerical methods described above. Eq. (2) is then solved to obtain the acceleration in the *o-xyz* frame. The velocity and the position of the ship are integrated by using the 4th-order Runge–Kutta method according to the following vector form rewritten from Eq. (2):

$$\begin{aligned}
\boldsymbol{\nu}_{k+1} &= \boldsymbol{\nu}_k + \frac{1}{6}(\mathbf{k}_1 + 2\mathbf{k}_2 + 2\mathbf{k}_3 + \mathbf{k}_4) \\
\mathbf{k}_1 &= f(\boldsymbol{\nu}_k, \mathbf{f}_k)dt \\
\mathbf{k}_2 &= f\left(\boldsymbol{\nu}_k + \frac{\mathbf{k}_1}{2}, \mathbf{f}_k\right)dt \\
\mathbf{k}_3 &= f\left(\boldsymbol{\nu}_k + \frac{\mathbf{k}_2}{2}, \mathbf{f}_k\right)dt \\
\mathbf{k}_4 &= f(\boldsymbol{\nu}_k + \mathbf{k}_3, \mathbf{f}_k)dt
\end{aligned} \tag{38}$$

Where

$$\dot{\boldsymbol{\nu}}_k = f(\boldsymbol{\nu}_k, \mathbf{f}_k)$$

$$= \mathbf{A}^{-1}\mathbf{f}_k - \mathbf{A}^{-1}\mathbf{N}(\boldsymbol{\nu}_k)$$

$$\boldsymbol{\nu}_k = [u_k, v_k, r_k]^T$$

$$\mathbf{f}_k = \begin{bmatrix} X_H + X_P + X_R + X_{ice} \\ Y_H + Y_R + Y_{ice} \\ N_H + N_R + N_{ice} \end{bmatrix}$$

$$\mathbf{A} = \begin{bmatrix} m & 0 & 0 \\ 0 & m & mx_G \\ 0 & mx_G & I_z \end{bmatrix}$$

$$N(\boldsymbol{\nu}_k) = \begin{bmatrix} -mv_k r_k - mx_G r_k^2 \\ mu_k r_k \\ mx_G u_k r_k \end{bmatrix}$$

where the subscript k and $k + 1$ represent the current and the subsequent time steps, respectively. The position and the orientation of the ship are calculated by using Eq. (38) along with Eq. (1). Coordinates of the nodes on the ice edge are then updated based on the new position and orientation.

3. SIMULATION RESULTS

The R-Class icebreaker, CCGS Sir John Franklin, was used in the present studies. It has a relatively simple underwater hull form. The vessel is one of most referred icebreaking hull forms in the literature since it was built in 1979. Both full-scale sea trials and ship model tests at various scales of the ship have been carried out for this icebreaker (Liu, 2009). To validate the proposed method, a series of Planar Motion Mechanism (PMM) test results were used. The numerical predictions were compared with the results of full-scale sea trials involving straight motion and turning.

The principal dimensions of the ship and the ice properties at model scale are listed inTable 1 and Table 2, respectively. The waterline profile is presented in Fig. 8.

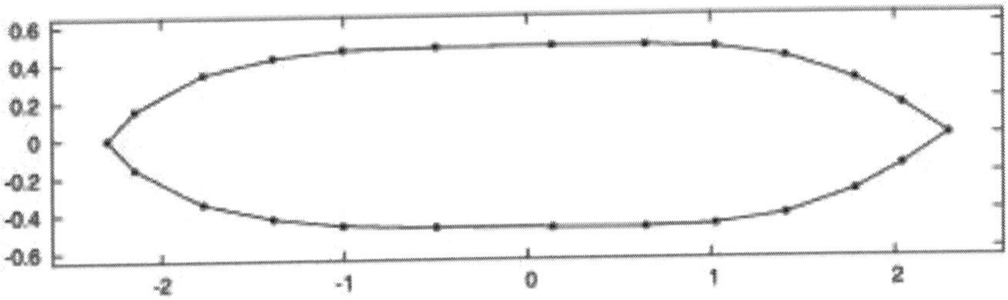

Figure 8. R-Class icebreaker ship model (model scale, unit: *m*).

Table 1. Principal dimensions (model scale).

Parameter	Notation	Value	Unit
Scale	λ	20	–
Length	L_{WL}	4.65	m
Breadth	B	0.96	m
Draft	T	0.35	m
Block Coefficient	C_B	0.611	–

Table 2. Ice mechanical properties (model scale).

Parameter	Notation	Value	Unit
Flexural strength	σ_f	20	kPa
Crushing strength	σ_c 100	kPa	
Ice thickness	h_i	28	mm
Young's modulus	E	28,140	kPa
Poisson ratio	ν	0.33	–
Friction coefficient	μ	0.1	–
Ice density	ρ_i	910	kg/m^3

3.1. Convergence studies

Prior to simulating the PMM tests, convergence studies were carried out to investigate the effect of the discrete length (the distance between two adjacent points) and the time step on the mean ice load. The applicability of the present method to ships at different scales and speeds were also studied. The 1:20 scale R-Class icebreaker model and the full-scale ship were used.

In the simulation of the straight tow tests, the chosen velocities were 0.5 m/s and 0.7 m/s in model scale.

The nondimensional discrete length, *l'*, and the nondimensional time step, *dt'*, were defined as follows:

$$l' = \frac{l}{B} \tag{39}$$

$$dt' = \frac{l}{u \cdot dt} \tag{40}$$

where *l* is the discrete length and *dt* is the time step used in the simulations.

Fig. 9 presents the mean ice resistances computed with various discrete lengths, andFig. 10 shows the resistances computed using different time steps. Good convergence in model scale is observed. The corresponding speeds for full scale are 2.236 m/s (4 knots) and 3.13 m/s (6 knots). Since the full-scale measurements are not available at these speeds, the simulated results were compared to the empirical solution based on the work of Spencer and Jones (2001). The resistance in full scale was estimated to be 389.6 kN and 442.5 kN, respectively. Good agreement is also found for full scale cases.

From Fig. 9 and Fig. 10, it can be seen that the computations are converged in all the cases when the nondimensional discrete length is smaller than 0.01 and the nondimensional time step is greater than 5. This indicates that the scale and the ship speed will not affect the accuracy of simulation if proper values of nondimensional length and time step are selected, i.e., the proposed numerical method is suitable for various scales and ship speeds. This conclusion needs to be further verified by using experimental and/or numerical results in different scales. In the following simulations, the nondimensional discrete length and time step were chosen as 0.005 and 5, respectively.

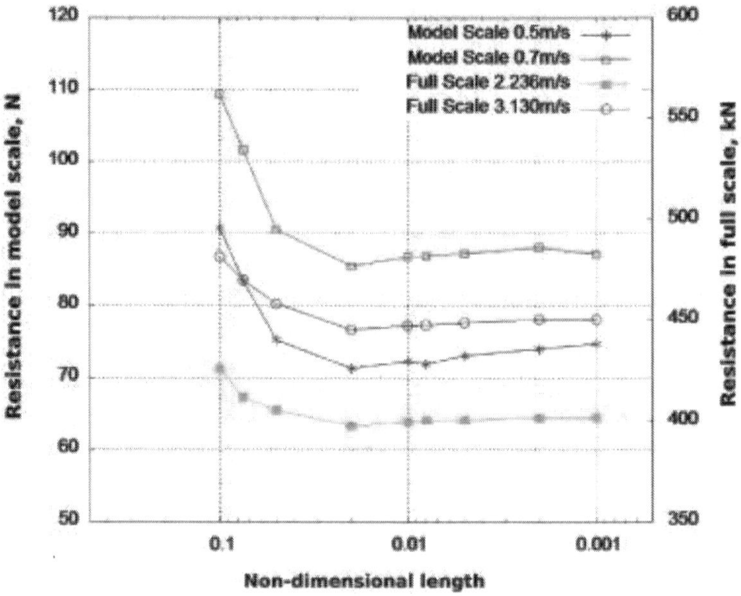

Figure 9. Convergence of ice resistance with respect to the discrete length.

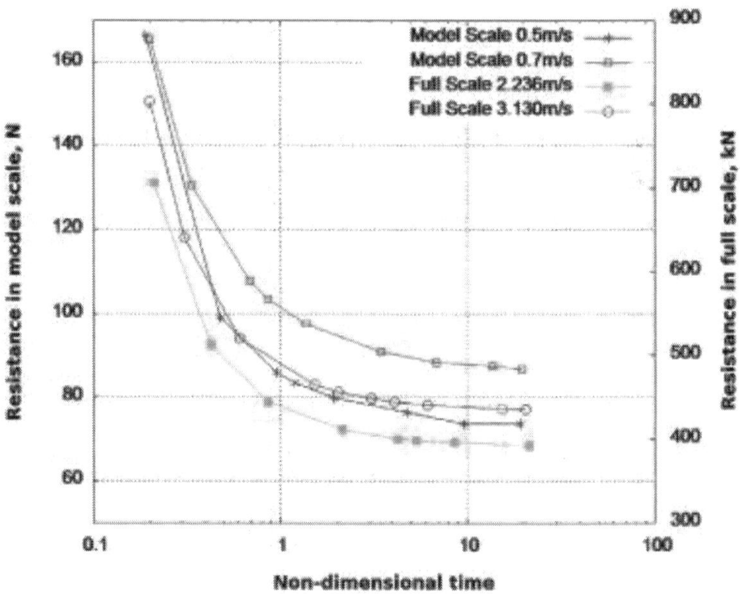

Figure 10. Convergence of ice resistance with respect to the time step.

3.2. Study of empirical parameters

The parameters, *Cf*, *Cl* and *Cv*, need to be determined for different types of ice and vessels. Note that *Cf* is used to determine the peak force of a single contact and *Cl* and *Cv* are used to determine the size of a broken ice piece. The parameter, *Cl*, significantly affects the time history of ice loads since smaller ice pieces will be broken off if a ship rapidly collides with the ice sheet.

The first step is to select *Cl* according to physical observations. Wang (2001) used 0.32 in her study; Lau et al. (2004) used a different model but the crack to be 0.2 of the characteristic length. According to the observation from Liu et al. (2008), *Cl* is selected to be 0.3 in this study. Once *Cl* is determined, the values of *Cf* and *Cv* are tuned by comparing the average resistance of simulation results against experimental measurements. Fig. 11 presents the mean resistance against ship velocity with different *Cf* values. Good agreement is observed between measurements and simulations when *Cf* is 2.2. The parameter, *Cv*, is used to include the speed effect on the size of the broken ice floe. Wang (2001) used − 0.14 for *Cv* in her study; Lau et al. (2004) does not include the velocity effect when determining the size of the crack cusp. *Cv* is set as − 0.5 in model scale. It should be pointed out that *Cv* has unit of *s/m*, and Froude's law is applied to get the value for full scale ($Cv = -0.11 s/m$).

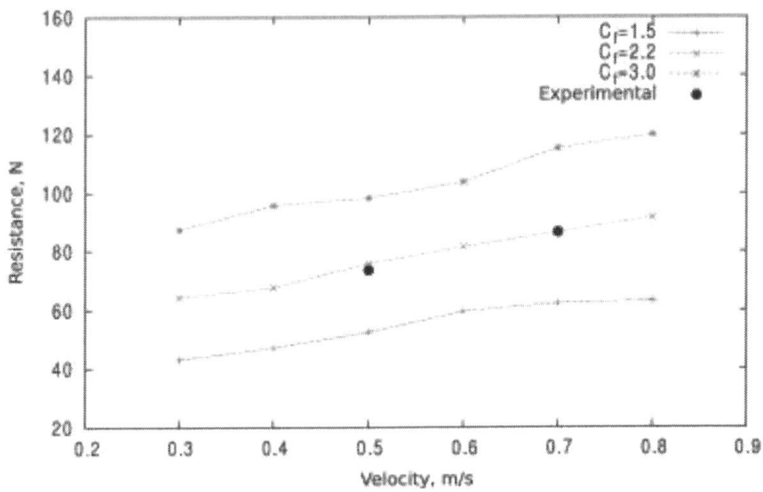

Figure 11. Resistance versus velocity with different C_f.

As shown in Fig. 11, the change of *Cf* will affect the relationship between the speed and the resistance (e.g., larger values of *Cf* result in a more steep increase trend of resistance). Once the value is determined, it can be used for all the simulations. A larger*Cf* results in a larger peak value and hence a larger mean resistance.

Note that *Cl* denotes the proportion of broken ice radius and the characteristic length as shown in Eq. (17). With a smaller value of *Cl*, smaller ice floe will be generated, which results in more frequent interaction between the vessel and the ice. This manifests into a larger mean ice resistance which is shown in Fig. 12.

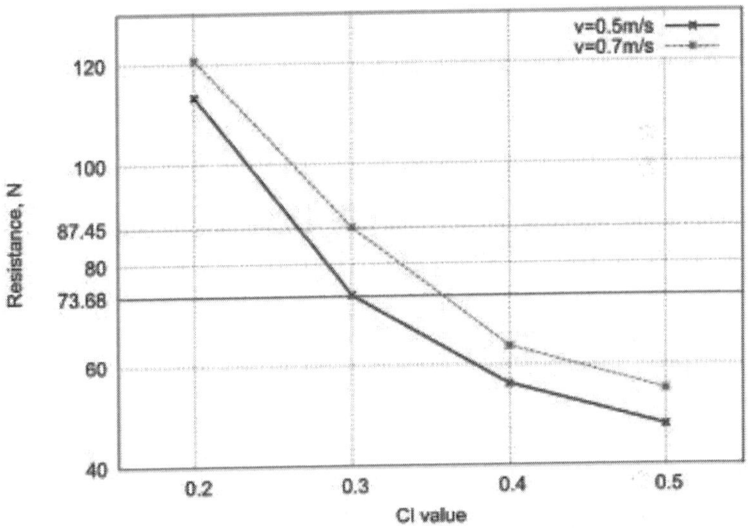

Figure 12. Resistance versus C_l with different velocities.

Cv illustrates the size of ice floe varies according to normal velocity of collision. A higher velocity of the vessel results in a higher normal velocity of collision and hence a smaller ice floe. Fig. 13 presents the histogram of ratio of crack radius and ice thickness at varying velocities. The values of radius fall within a limited range. There is no obvious trend that radius distribution changes according to vessel velocities. Fig. 14 gives the relationship between mean resistance and velocity with different *Cv* values. Linear relationship is observed in all cases, but the slopes are different.

Smaller value of *Cv* results in larger resistance. However, comparing to *Cf* and *Cl*, the effect of *Cv* on resistance is limited.

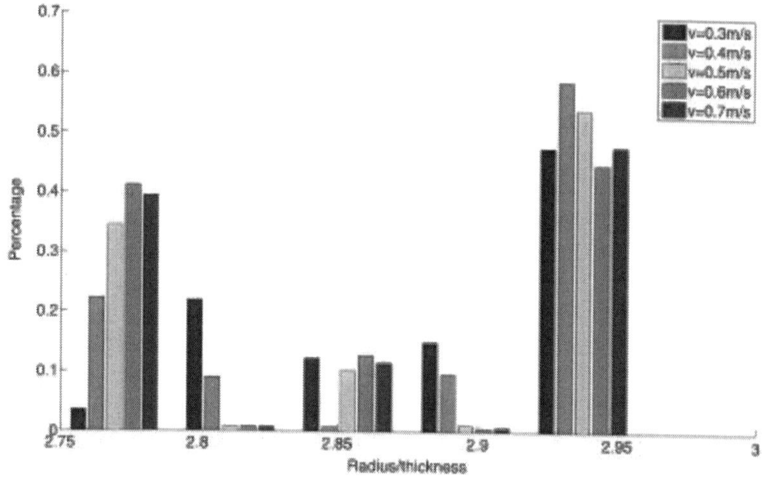

Figure 13. Ratio of crack radius and ice thickness at varying velocities.

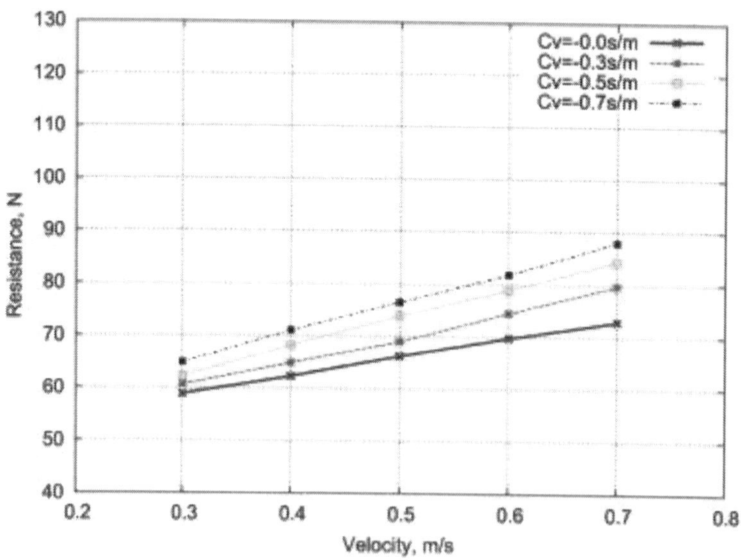

Figure 14. Resistance versus velocities with different C_v.

All three parameters will affect the mean resistance. *Cf* determines the peak value of the breaking force and directly affects the mean resistance; *Cl* and *Cv* directly determine the size of ice floe and have indirect effects on the resistance. Further studies are recommended for those parameters if the measured peak value and ice floe sizes are available from field or model test data.

3.3. Effect of flexural ice plate model

A flexural ice plate model was introduced to solve the ship-level ice interaction problem. The effect and the significance of the model were studied in model scale. A single crushing-bending- breaking event was simulated, which started from the ship stem contacting with a semi-infinite ice plate until an ice floe was broken off the plate. Three different ice plates were used: a rigid ice plate, a flexural ice plate with a high flexural strength ratio ($E/\sigma f = 6400$), and an ice plate with a low flexural strength ratio ($E/\sigma f = 1400$). The strength ratios were determined based on the work of Valanto (1989).Fig. 15 presents the breaking force versus the simulation time. When the ship contacts with the ice plate, the breaking force increases from zero to a maximum value where the breaking failure occurs. After that, the ship loses contact with ice and the breaking force falls back to zero.

It is clear that the maximum values of breaking forces are identical ($-20.90N$) in the three cases; however, the interaction durations and the mean forces vary during the interactions. Fig. 15 indicates that the lower the flexural strength ratio of the ice sheet is, the longer the duration of interaction will be. Table 3 presents a comparison between different types of ice sheets. The duration and the maximum vertical deformation increase greatly, which indicates bending is an component of this single event. However, a very small difference was observed in the mean breaking forces, especially for the ice floes with different flexural strength ratios.

The deflection ratios in the table represent the ratios between the flexural deflection and the characteristic length of the ice plate. Those values are smaller than 0.1, which verifies that the assumption of parallel downward movement is valid. The assumption that neglects the ship vertical movement can also be verified by comparing the maximum breaking force with the ship displacement.

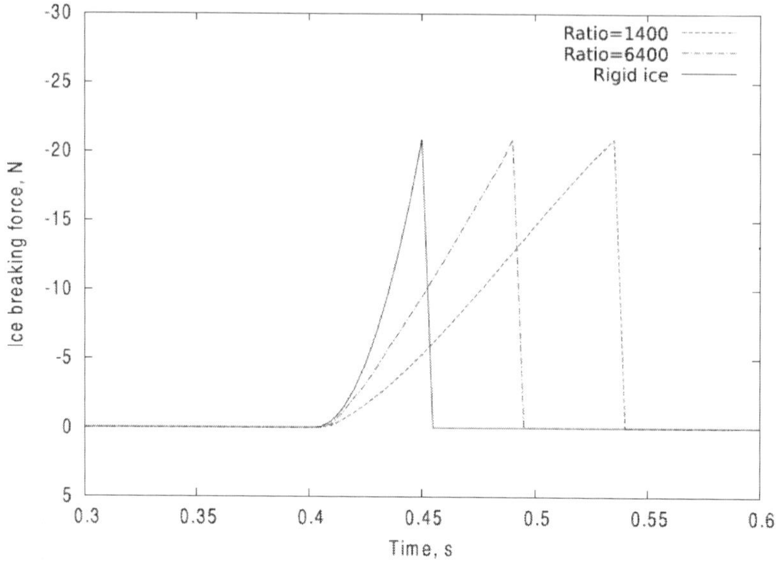

Figure 15. Breaking force in a single crushing-breaking event (0.5 m/s).

Table 3. Parameters for a single crushing-bending-breaking event.

Ice type	δ_v, $10^{-2}m$	δ_e/δ_v	Duration, s	δ_e/l	X_{br}, N
Rigid ice	1.295	0.00%	0.050	0.000	−7.609
High ratio	2.343	44.92%	0.090	0.032	−9.260
Low ratio	3.553	63.67%	0.135	0.069	−9.493

3.4. Model-scale PMM tests

PMM tests were simulated and compared with model test results to validate the numerical predictions. In the PMM tests, the amplitude of the lateral displacement was 2.5 m and the ice loads in surge, sway and yaw were measured. Average loads were recorded every 2 seconds. Each test lasted 50s. In order to compare the simulation results with the measured data, the identical test setup and data processing technique were applied in the simulation. A sample of ship model track and the broken ice channel is shown inFig. 16.

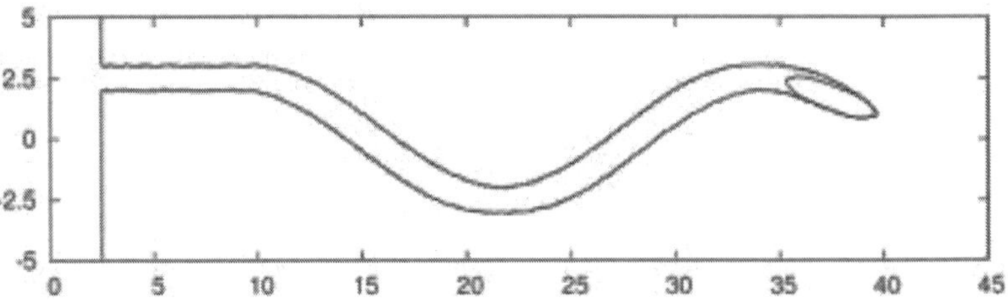

Figure 16. Broken ice channel in a pure yaw model test (unit: *m*).

Fig. 17, Fig. 18 and Fig. 19 present the time histories of surge resistance, sway force, and yaw moment, respectively. The red lines represent the measured data and the blue scattered dots represent the simulation results. From Fig. 17, it can be observed that the measured surge resistance oscillates with a constant period (approximately 25 s) which is half of the set value of the test. This is because the sway velocity was not strictly zero in the model tests. The average measured resistance is 73.680 N. A consistent average resistance (73.677 N) was obtained from the simulation. However, the explicit periodicity cannot be observed in the numerical results.

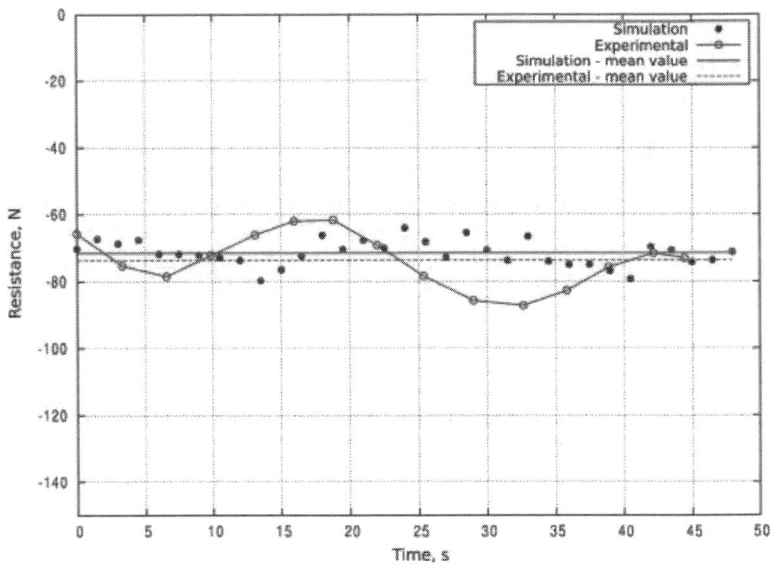

Figure 17. Time history of surge resistance.

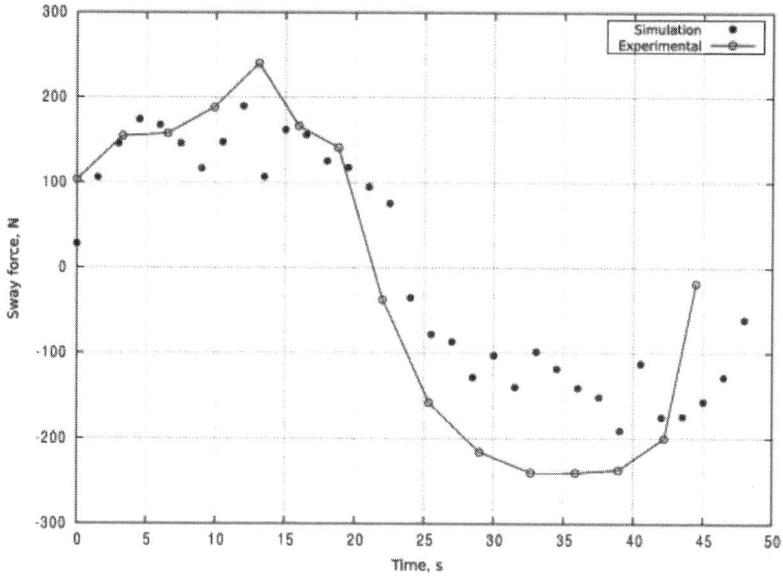

Figure 18. Time history of sway force.

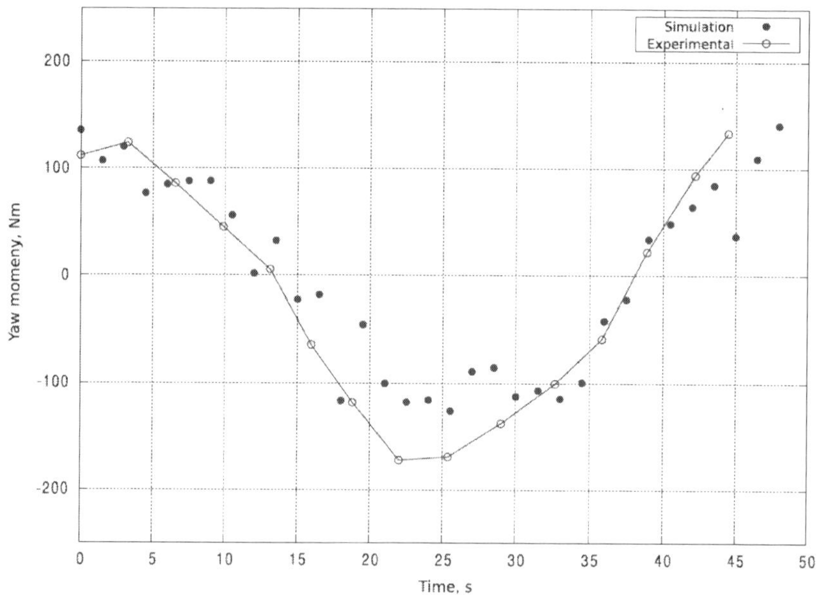

Figure 19. Time history of yaw moment.

In Fig. 18 and Fig. 19, the consistent and explicit oscillatory motions can be seen in both the measured and numerical results. It is also clear that the amplitudes of computed sway force and yaw moment are slightly smaller than the measured data. This phenomenon is validated by the regression relationship as shown in Fig. 20. One possible factor that results in this phenomenon is the non-zero sway velocity. Due to the existence of sway velocity in the pure yaw test, the mean force acting on midship may be estimated incorrectly so that the predicted sway force and yaw moment are smaller than measured values. This should be further investigated with more experimental data.

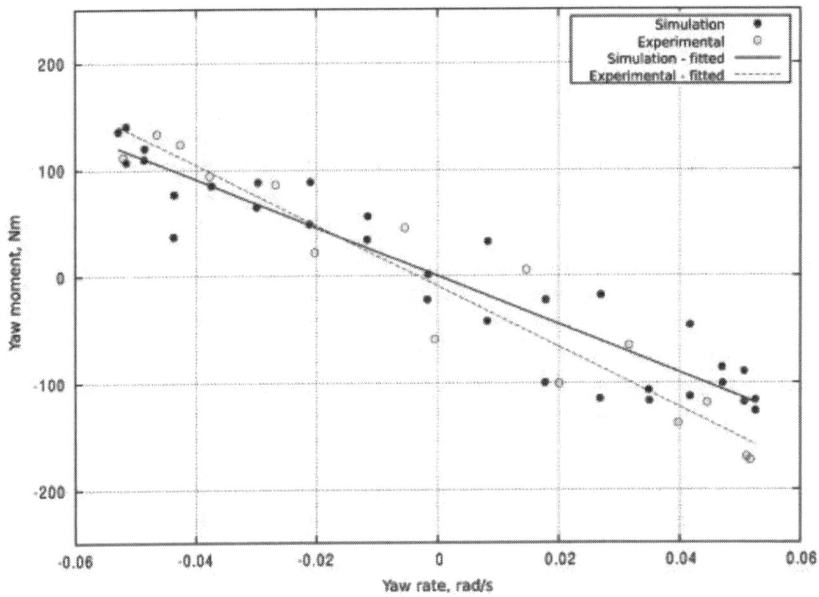

Figure 20. Relationship between yaw moment and yaw rate.

3.5. Full-scale straight running test

The full-scale results of free running tests were used to validate the total resistance and the propulsive performance of the ship. The full-scale free running tests were simulated in open water and in level ice. The computed propeller thrust and the total resistance were compared to the sea trial data from Keinonen (1996) and the analytical solutions using the formulas from Spencer and Jones (2001). The speed range of the tests in open water is from 6.34 m/s to 8.35 m/s, and that of the test in level ice is from 2.3 m/s to 6.3 m/s. The ice thicknesses vary from 0.489 m to 0.592 m. The simulations were conducted by restricting sway and yaw motions of the ship. Comparisons were made among the numerical results, empirical solutions and the sea-trial data.

Fig. 21 and Fig. 22 present the resistance and the propeller thrust in open water. It can be observed that the computed open water resistance and the thrust follow the same trend of the sea trial results, but are slightly smaller.

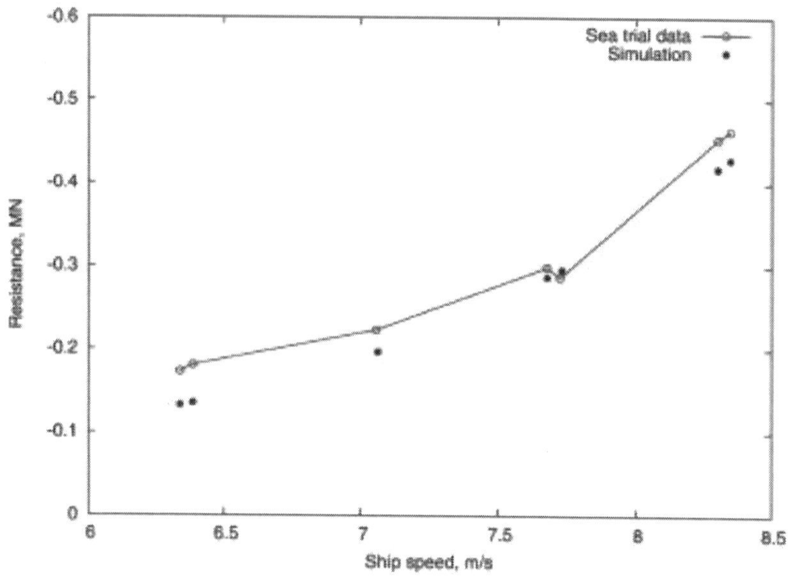

Figure 21. Resistance in open water.

The total resistance and the thrust in ice are presented in Fig. 23 and Fig. 24. The numerical results by the present method are compared with the sea-trial data and the empirical solutions. The computed resistances are in good agreement with the empirical results, while the propeller thrusts based on numerical simulations and empirical formulas are not in good agreement. Neither the empirical results of the thrust nor the predicted thrusts show good agreement with the sea-trial measurement.

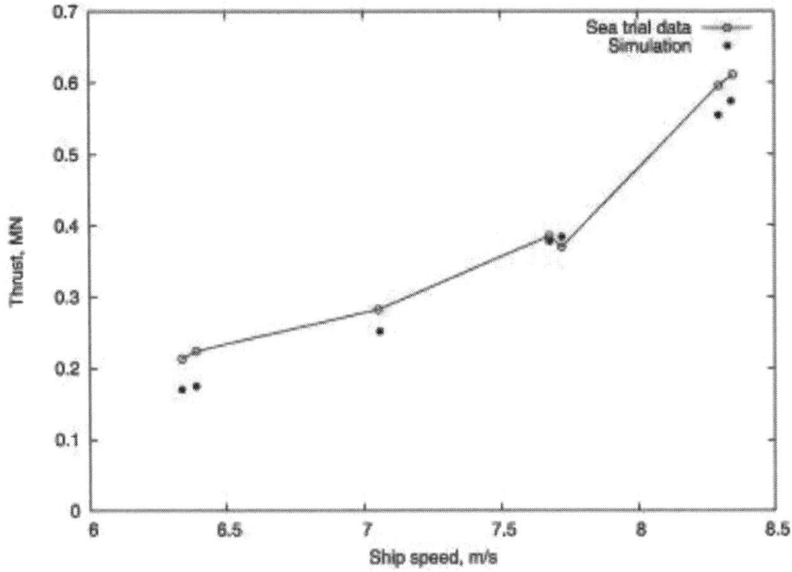

Figure 22. Thrust in open water.

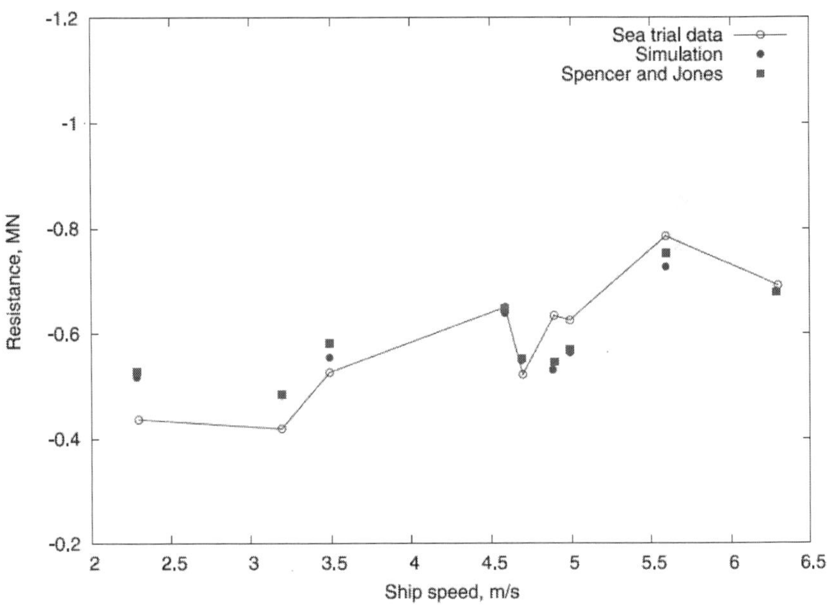

Figure 23. Resistance in level ice covered water.

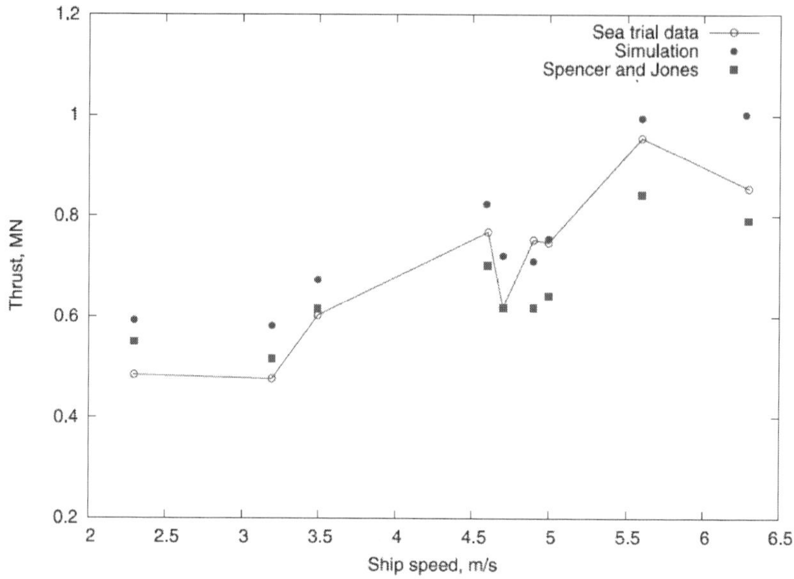

Figure 24. Thrust in level ice covered water.

The relatively poor agreement between the numerical prediction and the measurement is not necessarily a weakness of this model since the thrust is difficult to measure accurately in the field trials. The fluctuated nature of the predicted ice resistance may be responsible for the discrepancy between the predicted and empirical thrusts. From an energy perspective, the computed ice load oscillates with a relatively small mean value and high peaks. This may result in more energy consumption in order to maintain the speed. Therefore, a larger mean thrust was obtained from the simulation.

3.6. Full-scale turning test

The turning performance was also investigated and compared with the results of sea trials in open water and in level ice (Keinonen, 1996). All trials were conducted with 35°starboard rudder. The comparison of the steady speed in turn and the turning diameter between sea trials and the simulation results is presented in Table 4. The turning circles and the ice breaking channel are presented in Fig. 25. The sample time histories of ship velocity and ice load in surge, sway, and yaw in the ice field are presented in Fig. 26 and Fig. 27.

Table 4. Full-scale turning performance, $L_{WL} = 92.31$ m.

Condition	Shaft resolution	Speed in turn, *m/s*		Turning diameter, D/L_{WL}	
	RPM	Sea trial	Simulation	Sea trial	Simulation
Open water	120	3.8	3.9	4.21	4.1
Level ice	170	5.3	5.14	13.12	12.80

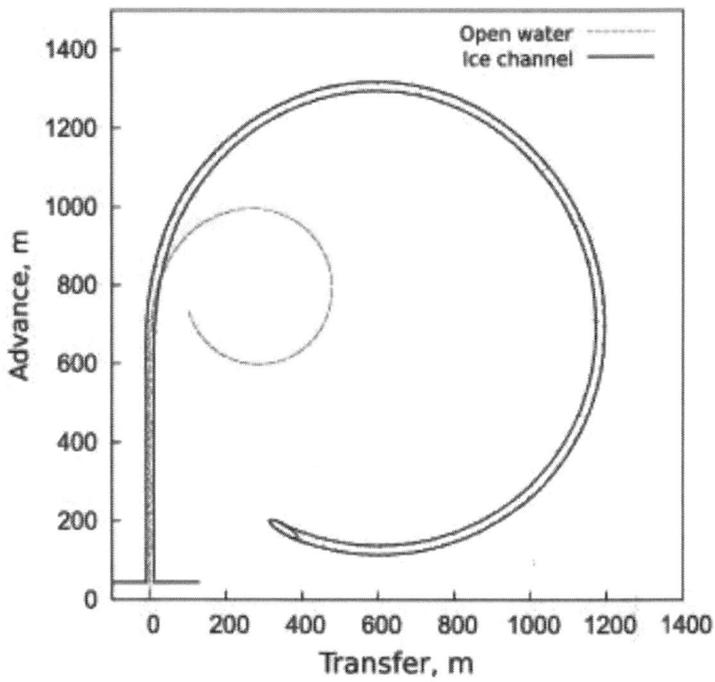

Figure 25. Turning track and ice channel.

In the simulations, the ship was kept straight for 600 m and then was turned using a full rudder angle (35°). The initial speed was 6.17 m/s (12.0knots). The shaft speed was 170 rpm in ice and 120 rpm in water. Good agreement between the sea-trial and simulation results indicates that the numerical method is able to simulate the turning of the ship. The turning circles are shown in Fig. 25.

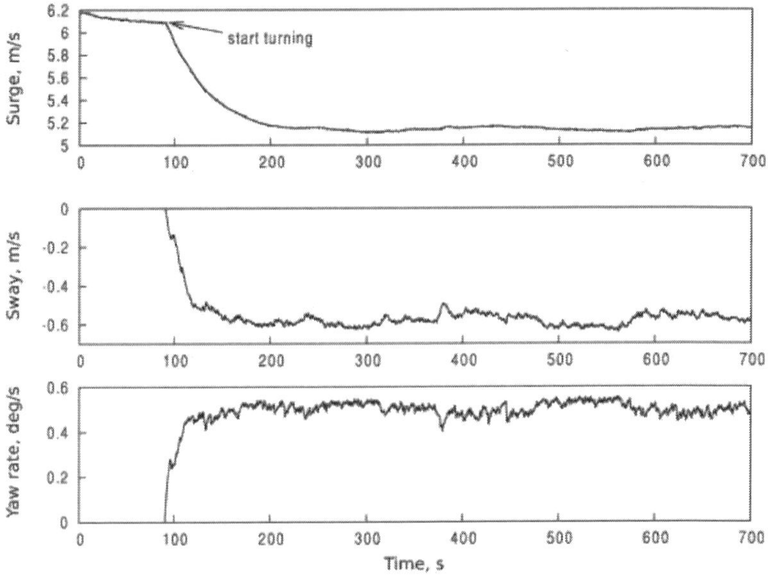

Figure 26. Time history of velocities (full scale).

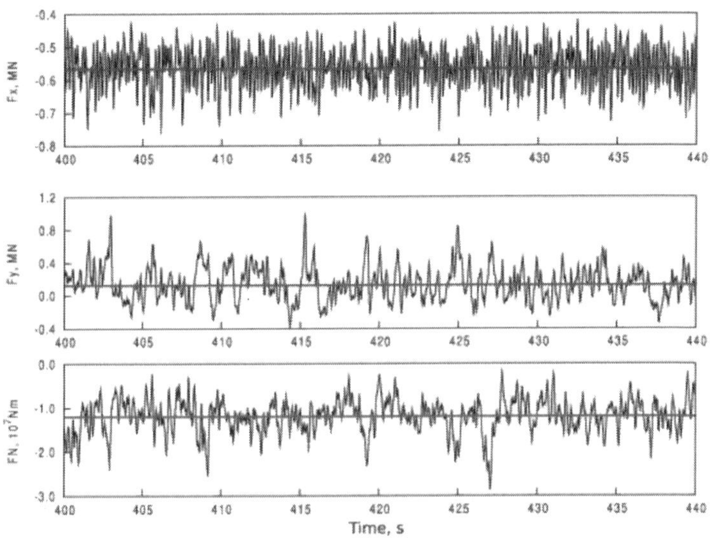

Figure 27. Time history of ice loads and mean values (full scale).

Fig. 26 presents the 3DOF ship velocities during the simulation period (700 sec). Fig. 27shows the 3DOF ice loads in a time period of 40 s during turning in the level ice. The ship started turning at 90.48 s and obtained a stable speed (5.14 m/s) at 200 s. The steady sway speed is 0.58 m/s to the port side and the yaw rate is 0.5 deg/s clockwise. The mean resistance during the steady turning is 0.566 MN. The mean value of yaw moment is 11.93 MNm. The velocities and the ice loads in sway and yaw each have non-zero value due to the turning. They oscillate around the mean value due to the ice breaking process.

4. CONCLUSIONS

A numerical model for simulating ship maneuvering performance in level ice has been developed and verified by comparing the numerical results to measured data and empirical solutions of the R-Class icebreaker, CCG Sir John Franklin. The numerical model has been developed considering similar strategies and assumptions made by previous researchers. It has been extended by considering the effect of buoyancy and clearing terms on sway and yaw motions. Moreover, the effect of the flexural deflection of ice plate is considered in the model. The water support of the ice plate is treated as an elastic foundation. The main conclusions can be drawn as follows:

1 A contact detecting method (the polygon-point algorithm) was proposed. As shown inFig. 5 and Fig. 6, good convergence was obtained in a wide range of discrete lengths and time steps. By using nondimensional length and time step, the effects of scaling and velocity were avoided. This indicates that the identical values of nondimensional discrete length and time step could apply to different cases.
2 The effect of the elastic foundation for the ice plate on ship–ice interaction process was considered. The elastic foundation tends to increase the mean ice resistance and extends the duration of each single contact-bending-breaking process.
3 A method was developed to consider the effect of buoyancy and clearing forces on sway and yaw motions. It is based on an empirical calculation of resistance and takes the hull geometry and the ship speed into account. It is validated by using the model PMM tests and the full-scale sea trial results. Therefore, the proposed numerical method is suitable for both model-scale and full-scale simulations.

ACKNOWLEDGEMENTS

The financial support from the NSERC CREATE Training Program for Offshore Technology Research (OTR) is highly appreciated. Mr. Don Spencer is gratefully acknowledged for his generous help on this research. The assistance by Dr. Claude Daley on the development of the flexural ice plate model is also acknowledged.

REFERENCES

1. Daley, C., 1991. Ice Edge Contact-A Brittle Failure Process Model. D.Sc. thesis, Acta Polytechnica Scandinavica, Mechanical Engineering Series No. 100. Helsinki.
2. Daley, C., 1992. Ice edge contact and failure. Cold Reg. Sci. Technol. 21, 1–23.
3. Daley, C., Tuhkuri, J., Riska, K., 1998. The role of discrete failures in local ice loads. Cold Reg. Sci. Technol. 27, 197–211.
4. Enkvist, E., 1972. On the ice resistance encountered by ships operating in the continuous mode of icebreaking. Technical Report 24 The Swedish Academy of Engineering Sci- ence in Finland.
5. Gong, I.-Y., 1993. Development of maneuvering simulation program, part I: mathematical models for maneuvering motions in deep/shallow water with wind/current effects. Technical Report IR-1993-09. Institute of Marine Dynamics, NRC.
6. Hirano, M., 1981. A practical calculation method of ship maneuvering motion at initial de- sign stage. Nav. Archit. Ocean Eng. 19, 68–80.
7. Holtrop, J., Mennen, G.G., 1982. An approximate power prediction method. Int. Shipbuild. Prog. 29, 166–170.
8. Kashteljan, V.I., Poznjak, I.I., Ryvlin, A.J., 1969. Ice Resistance To Motion Of A Ship. Marine Computer Application Corporation.
9. Keinonen, A., 1996. Icebreaker characteristics synthesis. Technical Report TP 12812E. AKAC Inc.
10. Kerr, A.D., 1976. The bearing capacity of floating ice plates subjected to static or quasi- static loads. J. Glaciol. 17, 229–268.
11. Kobayashi, H., Blok, J., Barr, R., Kim, Y., Nowicki, J., 2003. Specialist committee on Esso Osaka: final report and recommendations to the 23rd ITTC. Proceedings of the 23rd International Towing Tank Conference, pp. 8–14.

12. Lau, M., Liu, J.C., Derradji-Aouat, A., Williams, F.M., 2004. Preliminary results of ship ma- neuvering in ice experiments using a planar motion mechanism. Proceedings of the 17th International IAHR Symposium on Ice, pp. 479–487.
13. Lewis, J.W., Edwards Jr., R.Y., 1970. Methods For Predicting Icebreaking And Ice Resistance Characteristics Of Icebreakers. Transactions of Society of Naval Architects and Marine Engineers (SNAME), pp. 213–249.
14. Lindqvist, G., 1989. A straightforward method for calculation of ice resistance of ships. Proceedings of 10th International Conference on Port and Ocean Engineering under Arctic Conditions (POAC), pp. 722–735.
15. Liu, J., 2009. Mathematical Modeling Ice-Hull Interaction For Real Time Simulations Of Ship Manoeuvring In Level Ice (Ph.D. thesis) Memorial University of Newfoundland.
16. Liu, J., Lau, M., Williams, F., 2008. Numerical Implementation And Benchmark Of Ice-Hull Interaction Model For Ship Manoeuvring Simulations.
17. Martio, J., 2007. Numerical simulation of vessel's maneuvering performance in uniform ice. Technical Report Ship Laboratory. Helsinki University of Technology, Espoo, Finland.
18. Milano, V.R., 1972. Ship Resistance To Continuous Motion In Ice (Ph.D. thesis) Stevens Institute of Technology.
19. Nguyen, D.T., Sørbø, A.H., Sørensen, A.J., 2009. Modelling and control for dynamic posi- tioned vessels in level ice. Proceedings of 8th Conference of Manoeuvring and Control of Marine Craft (MCMC'2009), pp. 16–18.
20. Riska, K., Wilhelmson, M., Englund, K., Leiviska, T., 1997. Performance of merchant vessels in ice in the Baltic. Technical Report 52. Helsinki University of Technology - Ship Laboratory.
21. Sawamura, J., Riska, K., Moan, T., 2009. Numerical simulation of breaking patterns in level ice at ship's bow. Proceedings of 19th International Offshore and Polar Engineering Conference (ISOPE), pp. 600–607.
22. Spencer, D., 1992. A standard method for the conduct and analysis of ice resistance model tests. Technical Report IR-1992-12. Institute for Ocean Technology (IOT), NRC.
23. Spencer, D., Jones, S.J., 2001. Model-scale/full-scale correlation in open water and ice for Canadian coast guard "r-class": icebreakers. J. Ship Res. 45, 249–261.

24. Su, B., Riska, K., Moan, T., 2010. A numerical method for the prediction of ship perfor- mance in level ice. Cold Reg. Sci. Technol. 60, 177–188.
25. Tan, X., Su, B., Riska, K., Moan, T., 2013. A six-degrees-of-freedom numerical model for level ice–ship interaction. Cold Reg. Sci. Technol. 92, 1–16.
26. Tan, X., Riska, K., Moan, T., 2014. Effect of dynamic bending of level ice on ship's continuous-mode icebreaking. Cold Reg. Sci. Technol. 106, 82–95.
27. Valanto, P., 1989. Experimental study of the icebreaking cycle in 2-d. Proceedings of the 8th International Conference on Ocean. Offshore and Arctic Engineering (OMAE).
28. Wang, S., 2001. A dynamic model for breaking pattern of level ice by conical structures. Acta Polytechnica Scandinavica Mechanical Engineering Me 156.
29. Zhou, Q., Peng, H., 2014. Numerical simulation of a dynamically controlled ship in level ice. Int. J. Offshore Polar Eng. 24, 184–191.

CHAPTER 3

Numerical simulation for the free surface flow around a complex ship hull form at different Froude numbers

Y.M. Ahmed

Naval Architecture and Marine Engineering Department,
Alexandria University, Alexandria, Egypt

ABSTRACT

The incompressible turbulent free surface flow around the complex hull form of the DTMB 5415 model at two different speeds has been numerically simulated using the RANSE code CFX. The Volume of Fluid method (VOF) has been used with CFX for capturing the free surface flow around the ship model at the two speeds. The simulation conditions are the ones for which experimental and numerical results exist. The standard *k–ε* turbulence model has been used in CFX code. The grid generator ICEM CFD has been used for building the hybrid grid for the RANSE code solver. The results compare well with the available experimental and numerical data.

KEYWORDS

DTMB; CFD; Free surface; Wave; CFX; Resistance; Numerical

1. INTRODUCTION

Studying the pattern of waves generated by a ship moving through the water is one of the most important objectives in ship hydrodynamics, due to its importance in the design process. The waves produced by a ship in motion can radiate at great distances away from the ship. Furthermore, these waves contain energy that must be dissipated to the surrounding fluid. The ship experiences an opposing force to its movement; one of its components is known as the wave making resistance, this being one of the most important components of the ship resistance.

Potential and turbulent flow methods are being intensively used for investigating and simulating the flow pattern around ships. Typically, the potential flow codes based on the boundary element method (BEM) [1] and [2] are commonly used to study the ship waves on the free surface and to investigate the interaction between bulbous bow and forward shoulder waves. The effects of fluid viscosity are neglected in this case and the flow computation is relatively simple, quick, and accurate. The ship hull and the water surface in the potential flow codes are usually discretized using triangular or quadrilateral panels; hence the flow is computed by solving the Laplace equation.

For simulating wave resistance problems in potential flow codes, the Rankine source method [3] and [4] is considered to be the most widely used technique in this case. The free surface boundary condition can be incorporated successfully in the simulation process based on a so-called double model solution, which was proposed by Dawson [5]. Since then it has been widely applied as a practical method and many improvements have been made to account for the nonlinearity of the free surface physics [6].

Modern viscous flow codes, which solve the RANSE, have the ability to simulate efficiently the turbulent flow problems around ships with and without free surface effects. These codes can produce relatively accurate results both for the flow field and for the ship resistance. Furthermore, the use of the viscous codes is essential for investigating some flow characteristics such as flow separation near the ship stern, where the flow pattern can only be predicted reasonably by viscous approaches. The numerical methods are used for solving governing fluid equations with these codes, through discretization schemes such as finite differences and finite volume methods.

Free surface problems such as the wave making problem can be simulated with viscous flow codes depending on whether the computational grid adapts to the shape or the position of the free surface. Therefore, two major approaches widely applied to the free surface computations are the so-called interface-tracking method, e.g., a moving mesh[7] and [8], and the interface-capturing method, e.g., the volume of fluid method (VOF)[9] and [10].

The mesh in the former method moves over an underlying fixed Eulerian grid for tracking a free surface flow pattern around ship hull. This mesh only covers the domain involving the water, where the free surface forms the upper boundary of the computational domain and is determined as part of the solution. This approach can be applied for moving boundary problems, but special treatments are required for simulating problems of large deformation such as breaking waves.

In the interface-capturing method, both air and water are considered in the simulation and treated as two effective fluids. The numerical grids in this method are fixed in space, and the predication of the free surface location is achieved by solving an additional transport equation.

In this study, the incompressible free surface flow around the DTMB 5415 model has been studied using the finite volume flow code CFX [11] at $F_n = 0.28$ and 0.41. The frictional resistance and the pressure resistance (the viscous pressure resistance plus the wave making resistance) coefficients of the model hull have been calculated numerically using the finite volume code.

The flow was taken to be steady in the RANSE code simulations. The standard k–ε turbulence model was used in CFX simulations. The predicted numerical results were compared with the available experimental and numerical results for the complex hull form of the DTMB 5415 model at the two speeds.

2. DESCRIPTION OF DTMB 5415 HULL FORM

The model 5415 (Fig. 1) was conceived as a preliminary design for a Navy surface combatant ca. 1980. The hull geometry includes both a sonar dome and transom stern. Propulsion is provided through twin open-water propellers driven by shafts supported by struts.

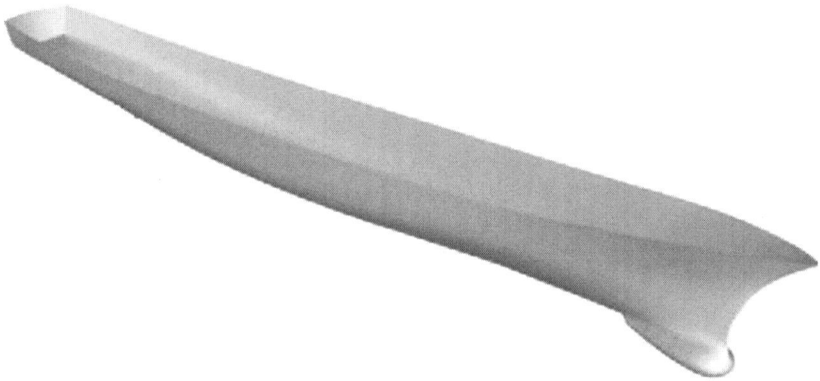

Figure 1. The hull form of DTMB 5415 unit.

The principal dimensions of the DTMB 5415 are listed in Table 1 and the test conditions that are listed in Table 2 correspond to the model tests from the David Taylor Model Basin (DTMB) in Washington DC and the Istituto Nazionale per Studied Esperienze di Architettura Navale (INSEAN) in Rome, Italy. Both computations and experiments are conducted in the bare hull condition.

Table 1. Principal dimensions of DTMB 5415 [12].

Length	142.037 m
Beam	17.983 m
Draft	6.179 m
Wet surface area	2976.7 m^2
Water plane area	1987.2 m^2
Displacement	12901.6 m^3
Block coefficient	0.506

Table 2. Test conditions for DTMB 5415 model scale [12].

Scale ratio	24.832
Length (*L*)	5.72 m
Draft (*T*)	0.248 m
Wet surface area	4.861 m^2
Froude numbers	0.28 and 0.41

3. MATHEMATICAL MODEL

The mathematical description of the free surface flow in CFX is based on the homogenous multiphase Eulerian–Eulerian fluid approach. Both fluids (water and air) in this approach share the same velocity field and other relevant fields such as temperature, turbulence, etc., and they are separated by a distinct resolvable interface. The local equations governing the motion of an unsteady, viscous, incompressible fluid (either liquid or gas) are the Navier–Stokes equations, which in a conservative formulation are given as:

$$\frac{\partial}{\partial t}(\rho) + \frac{\partial}{\partial x_i}(\rho u_i) = 0.0 \qquad (1)$$

$$\frac{\partial}{\partial t}(\rho u_i) + \frac{\partial}{\partial x_i}(\rho u_i u_j) = -\frac{\partial p}{\partial x_i} + \frac{\partial}{\partial x_i}(-\rho \overline{u_i' u_j'}) + \frac{\partial}{\partial x_j}\left[\mu\left(\frac{\partial u_i}{\partial x_j} + \frac{\partial u_j}{\partial x_i} - \frac{2}{3}\delta_{ij}\frac{\partial u_l}{\partial x_l}\right)\right] \quad (2)$$

where,

$$-\rho \overline{u_i' u_j'} = \mu_t\left(\frac{\partial u_i}{\partial x_j} + \frac{\partial u_j}{\partial x_i}\right) - \frac{2}{3}\left(\rho k + \mu_t \frac{\partial u_i}{\partial x_i}\right)\delta_{ij} \quad (3)$$

And

$$\rho = \sum_{\alpha=1}^{2} r_\alpha \rho_\alpha, \quad \mu = \sum_{\alpha=1}^{2} r_\alpha \mu_\alpha, \quad \sum_{\alpha=1}^{2} r_\alpha = 1$$

The standard k–ε turbulent model was used in CFX code, where

$$\rho u_j \frac{\partial k_i}{\partial x_j} = \frac{\partial}{\partial x_i}\left[\left(\mu + \frac{\mu_t}{\sigma_k}\right)\frac{\partial k}{\partial x_i}\right] + G_k + G_b - \rho\varepsilon - Y_M \quad (4)$$

$$\rho u_j \frac{\partial \varepsilon_i}{\partial x_j} = \frac{\partial}{\partial x_i}\left[\left(\mu + \frac{\mu_t}{\sigma_k}\right)\frac{\partial \varepsilon}{\partial x_i}\right] + C_{1\varepsilon}\frac{\varepsilon}{k}(G_k + C_{3\varepsilon}G_b) - C_{2\varepsilon}\rho\frac{\varepsilon^2}{k} \quad (5)$$

where G_k, G_b and Y_M are the generation of turbulent kinetic energy due the mean velocity gradients, the generation of turbulent kinetic energy due to buoyancy and the contribution of the fluctuating dilatation in compressible turbulence to the overall dissipation rate. In this study the model constants have taken the following values$C_{1\varepsilon} = 1.45$, $C_{2\varepsilon} = 1.9$, $\sigma_k = 1.0$ and $\sigma_\varepsilon = 1.3$.

4. COMPUTATIONAL GRID

The computational domain in this study extends for 1.5L in front of the ship hull, 2.5Lbehind the hull, 1.5L to the side and 1.2L under the keel of the model. The air layer extends 0.125L above the still water surface. The grid generator of the RANSE code (ICEM CFD) has been used for building the required hybrid mesh for the code solver. The computational domain of the DTMB 5415 model (Fig. 2) has been meshed with structured hexahedral grid of 95971 elements, while the hull surface of the DTMB 5415 model and the region around it have been meshed using unstructured tetrahedral grid of 331924 elements (Fig. 3). The mesh was refined in the free surface region in order to get a sharp free surface. Furthermore, for predicting accurate values for C_P the mesh elements were refined on and near the ship hull surface in order to calculate accurate pressure forces acting on the hull form at the two Froude numbers. Finally, the space of the first grid from the hull is $y^+ \approx 25$.

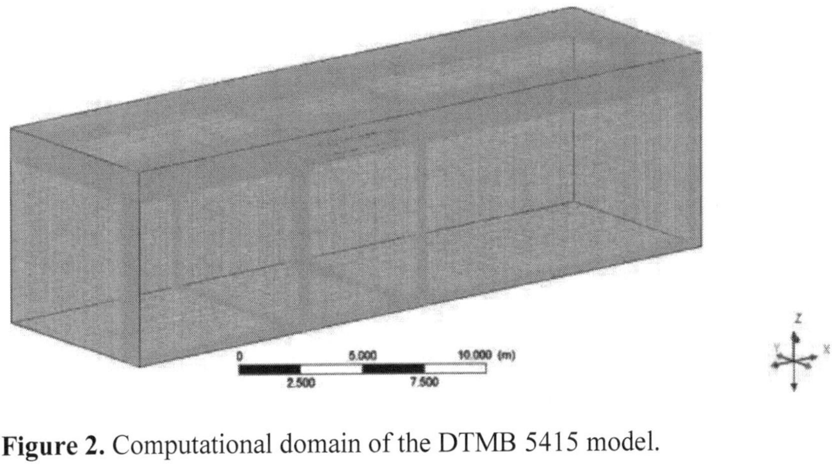

Figure 2. Computational domain of the DTMB 5415 model.

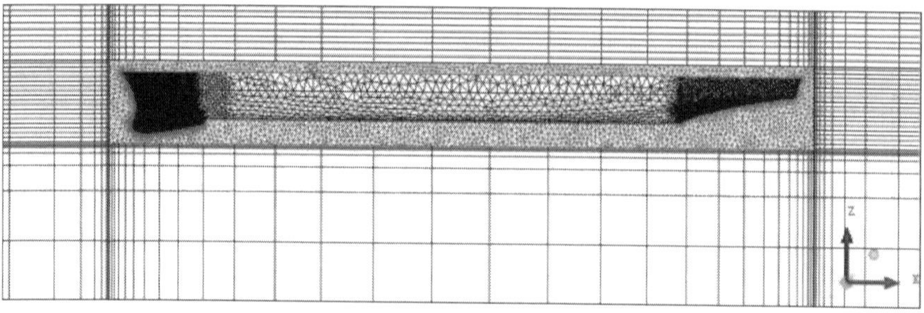

Figure 3. Hybrid mesh for the DTMB 5415 model and its domain.

5. BOUNDARY CONDITIONS AND COMPUTATIONAL METHODS

The no-slip boundary condition is imposed on the hull surface of the DTMB 5415 model, that is, the fluid particles on the body move with body velocity. The static pressure at the outlet boundary was defined as a function of water volume fraction. Furthermore, the initial location of the free surface was imposed by defining the volume fraction functions of water and air at the inlet and outlet boundaries. The scalable wall function was used with the turbulence model. The reference pressure was set to an atmospheric pressure. Only the starboard side of the DTMB 5415 model hull was considered with CFX due to the symmetric properties of the problem under consideration.

The flow was considered steady in CFX calculations, and the finite volume method is used for the discretization process. The high resolution numerical scheme was used for discretizing the advection terms. A linear interpolation scheme was used for interpolating the pressure, while the velocity was interpolated using a trilinear numerical scheme. The results in this study were obtained on one computer Pentium IV 2 GHz and 1 Gb RAM.

6. RESULTS

The wave cuts along the hull surface of the DTMB 5415 model at the two speeds are represented in Figure 4 and Figure 5. The comparisons between the predicted results and the available experimental results show in general good agreement between the results at the two speeds. However, the wave cut predicted by CFX at $F_n = 0.41$ is somewhat under predicted than that obtained from the experimental results, especially from nearly -0.3*L* to 0.5*L*. This may give a direct indication about the requirements of more unstructured tetrahedral mesh elements around the DTMB 5415 hull, which could not be attained due to the limited computational power available.

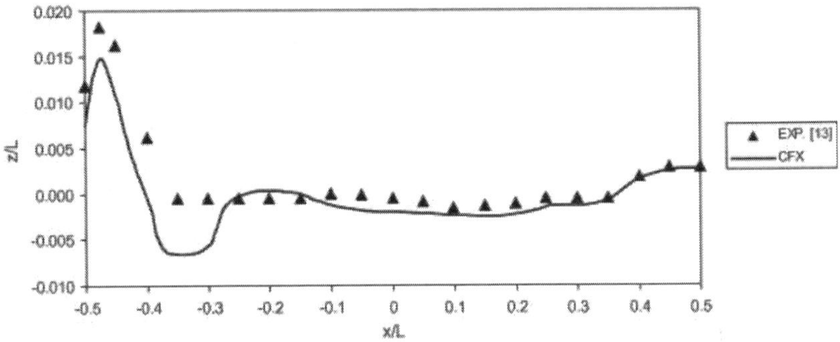

Figure 4. Wave cut along the hull of the DTMB 5415 model at $F_n = 0.28$.

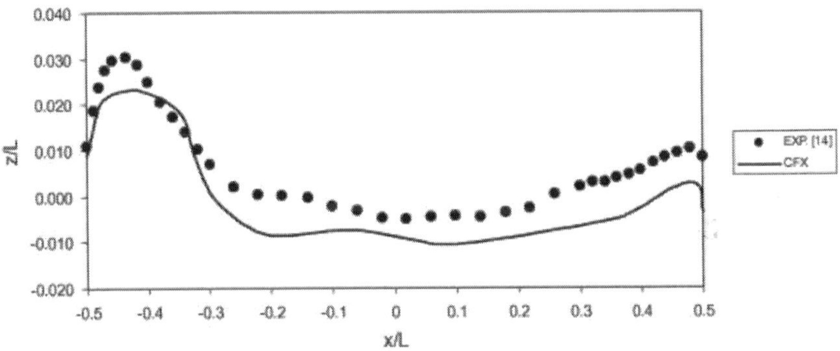

Figure 5. Wave cut along the hull of the DTMB 5415 model at $F_n = 0.41$.

The comparison between the predicted wave pattern around the hull form of the DTMB 5415 model by CFX, the available experimental results [15] and the numerical results obtained from using the potential flow method [16] at $F_n = 0.28$ (Fig. 6) shows the ability of CFX for predicting nearly the same wave contours at the bow and the stern of the model. However, the calculated wave contours at the mid region of the ship is different from the experimental and numerical contours at this speed. This may refer to the difficulty in calculating the wave contours at this low Froude number due to the small value of the wave making resistance in this case. On the other hand, there is a good agreement between CFX results and the numerical results [16] for the wave contours at $F_n = 0.41$ as can be seen from Fig. 7. The transverse waves and the complex shape of the sonar dome have a pronounced effect on the wave pattern in this case ($F_n = 0.41$), which has been predicted well by CFX.

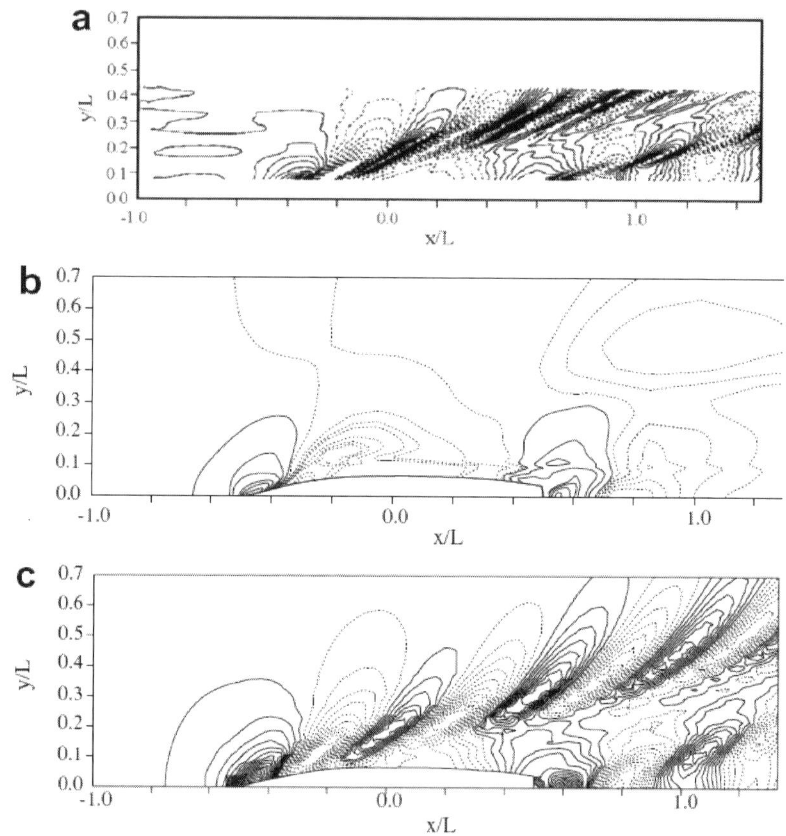

Figure 6. (a) Experimental wave contours at $F_n = 0.28$ [15], (b) predicted wave contours at $F_n = 0.28$ (CFX), (c) predicted wave contours by potential flow method at $F_n = 0.28$ [16].

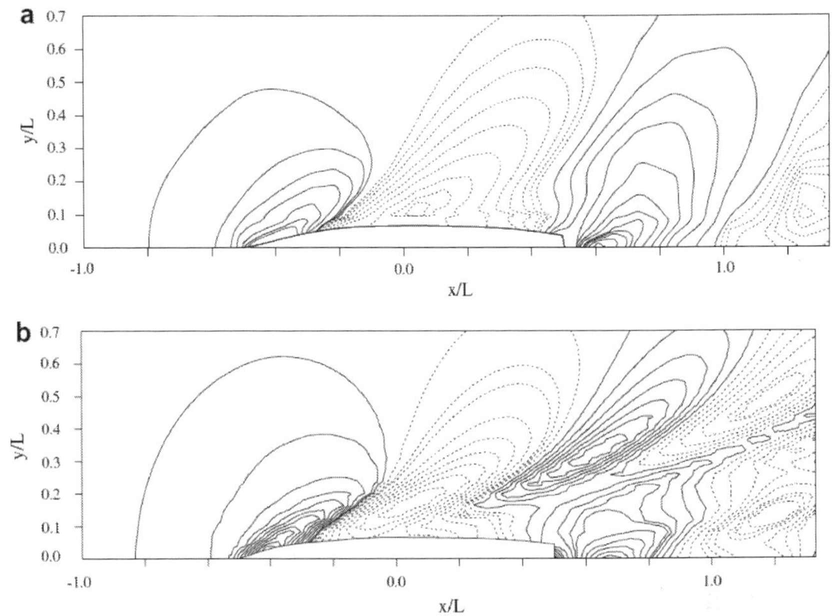

Figure 7. (a) Predicted wave contours at $F_n = 0.41$ (CFX), (b) predicted wave contours by potential flow method at$F_n = 0.41$ [16].

The detected distribution of the pressure coefficient on the model hull at $F_n = 0.28$ is presented in Fig. 8. The zones of high pressure at the bow and the aft regions are detected well by the code. The higher value of the pressure coefficient can be found at the stagnation region on the hull sonar dome. There is a pressure gradient aft of the stagnation pressure region and the pressure decreased at the bow and the sonar dome until it reached its lowest value as can be seen from the previous figure. The pressure increases again aft of the bow region due to the effect of hull geometry on the flow pattern around the hull at this speed. Downstream of this position the pressure again decreases, and approximately extends over the entire region of the hull mid part due to nearly the constant cross-section of the hull in this region. Finally, the effect of the hull wave pattern on the hull pressure distribution was detected well by the code at the draft region.

For the other speed ($F_n = 0.41$) the distribution of C_p on the hull is nearly the same as that of $F_n = 0.28$, as can be noted from Fig. 9. However, the pressure coefficient range at$F_n = 0.41$ has higher and lower values than that for $F_n = 0.28$. Moreover, the wave pattern at this speed has a more pronounced effect on the bow and the hull fore region pressure distribution, due to the increase of wave height in this case.

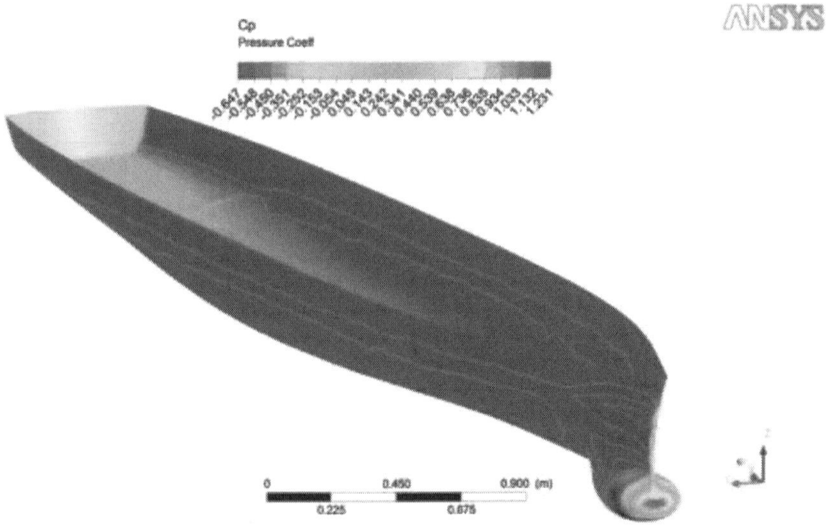

Figure 8. Distribution of C_p on the hull surface of DTMB 5415 model at $F_n = 0.28$.

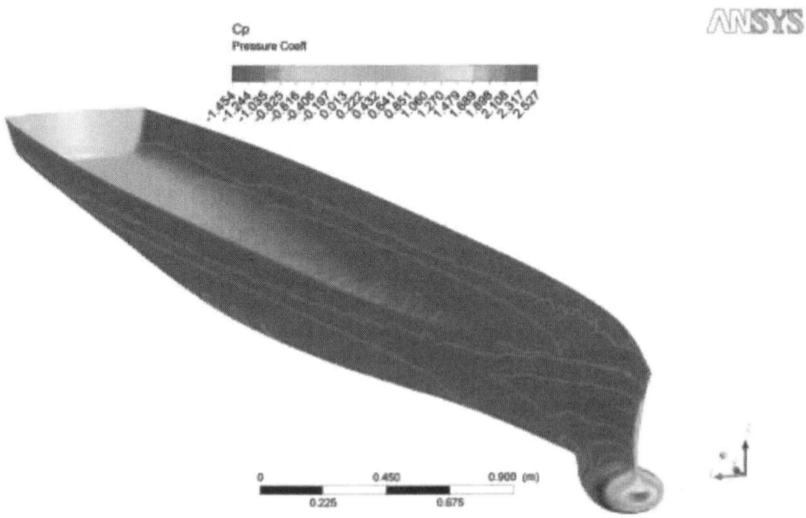

Figure 9. Distribution of C_p on the hull surface of DTMB 5415 model at $F_n = 0.41$.

The numerical results obtained for the pressure resistance coefficient, $C_P = R_P/0.5\rho A V_\infty^2$, and the frictional resistance coefficient, $C_F = R_F/0.5\rho A V_\infty^2$ by CFX at the two speeds, are presented in Table 3. The comparisons between the experimental results [17] and CFX results in the previous table show the ability of the RANSE code CFX in predicting accurate values for the resistance coefficients at the two speeds.

Table 3. Experimental and numerical resistance components of the DTMB 5415 model.

Froude no.	Coefficient	Experimental results [17]	CFX
$F_n = 0.28$	C_F	2.880×10^{-3}	2.995×10^{-3}
	C_P	1.350×10^{-3}	1.513×10^{-3}
	C_T	4.230×10^{-3}	4.384×10^{-3}
$F_n = 0.41$	C_F	2.806×10^{-3}	2.738×10^{-3}
	C_P	3.094×10^{-3}	2.919×10^{-3}
	C_T	5.900×10^{-3}	6.075×10^{-3}

7. CONCLUSIONS

The turbulent free surface flow around the DTMB 5415 model hull at two Froude numbers has been simulated using the RANSE code CFX. The wave profiles along the hull of the model compared well against the available experimental results at the two speeds.

The results of CFX code show the ability of predicting good results for the free surface wave pattern around the hull at $F_n = 0.41$, even with the need for more computational grid.

The most important conclusion that can be drawn from this study is the good benefit from using the hybrid mesh with a RNASE solver and limited computational resources for predicting the turbulent free surface flow pattern around a complex hull form such as that for the DTMB 5415 model without the use of very large number of computational grid elements [15] or the use of potential flow method for calculating the wave making resistance and wave patterns beside the use of viscous flow method for predicting the frictional resistance and the viscous pressure components of the model at the different speeds [16].

REFERENCES

1. J.L. Hess, Panel methods in computational fluid dynamics, Annual Review of Fluid Mechanics 22 (1990) 255–274.
2. J.L. Hess, A.M.O. Smith, Calculation of nonlinear potential flow about arbitrary three-dimensional bodies, Journal of Ship Research 8 (2) (1964) 22–44.
3. G. Jensen, H. Soding, Rankine methods for the solution of the steady wave resistance problem, in: Proceedings of the 16th Symposium on Naval Hydrodynamics, 1986, pp. 575–582.
4. L. Larsson, L. Broberg, K.J. Kim, D.H. Zhang, New viscous and inviscid CFD techniques for ship flow, in: Fifth International Conference on Numerical Ship Hydrodynamics, Hiroshima, Japan, 1989, pp. 1–25.
5. C.W. Dawson, A practical computer method for solving ship- wave problems, in: Proceedings of the 2nd International Conference on Numerical Ship Hydrodynamics, Berkeley, CA, 1977, pp. 30–38.
6. H. Raven, A solution method for the nonlinear ship wave resistance problem, Ph.D. thesis, Delft University, 1996.
7. G. Cowles, L.A. Martinelli, Viscous multiblock flow solver for free-surface calculations on complex geometries, in: Proceedings of the 22nd Symposium on Naval Hydrodynamics, 1998, pp. 9– 14.
8. T. Li, J. Matusiak, Simulation of modern surface ships with a wetted transom in a viscous flow, in: Proceedings of the 11th International Offshore and Polar Engineering Conference, vol. 4, 2001, pp. 570–576.
9. C.W. Hirt, B.D. Nichols, Volume of fluid (VOF) method for dynamics of free boundaries, Journal of Computational Physics 39 (1) (1981) 210–225.
10. V. Maronnier, M. Picasso, J. Rappaz, Numerical simulation of free surface flows, Journal of Computational Physics 155 (1999) 439–455.
11. Ansys CFX web page: <http://www.ansys.com/products/fluid-dynamics/cfx/> (accessed October 2010).
12. Y.J. Chen, S.W. Chau, J.S. Kouh, Application of two-phase fluid approach for free surface ship flow simulation, Journal of the Chinese Institute of Engineers 25 (2) (2002) 179–188.
13. L. Larsson, F. Stern, V. Bertram, Benchmarking of computational fluid dynamics for ship flows: the Gothenburg 2000 workshop, Journal of Ship Research 47 (1) (2000) 63–81.

14. Yasser, G.S. Carlos, Simulation of the flow around the surface combatant DTMB model 5415 at different speeds, in: 13th Congress of International Maritime Association of Mediterranean, Istanbul, Turkey, 2009, pp. 307–314.
15. DTMB web page: <http://www50.dt.navy.mil/5415/> (accessed April 2007).

CHAPTER 4

A framework for risk assessment for maritime transportation systems—A case study for open sea collisions involving RoPax vessels

Jakub Montewka [a], SörenEhlers b, FlorisGoerlandt [a], Tomasz Hinz [c], KristjanTabri [d], Pentti Kujala [a]

aAalto University,School of Engineering, Department of Applied Mechanics ,Research Group on Maritime Riskand Safety, P.O.Box15300, FI-00076 AALTO, Espoo, Finland

bNorwegian University of Science and Technology, Department of Marine Technology, Trondheim,Norway

c Waterborne Transport Innovation, Łapino, Poland

d Technical University of Tallinn ,Estonia

ABSTRACT

Maritime accidents involving ships carrying passengers may pose a high risk with respect to human casualties. For effective risk mitigation, an insight into the process of risk escalation is needed. This requires a proactive approach when it comes to risk modelling for maritime transportation systems. Most of the existing models are based on historical data on maritime accidents, and thus they can be considered reactive instead of proactive.

This paper introduces a systematic, transferable and proactive framework estimating the risk for maritime transportation systems, meeting the requirements stemming from the adopted formal definition of risk. The framework focuses on ship–ship collisions in the open sea, with a RoRo/Passenger ship (RoPax) being considered as the struck ship. First, it covers an identification of the events that follow a collision between two ships in the open sea, and, second, it evaluates the probabilities of these events, concluding by determining the severity of a collision. The risk framework is developed with the use of Bayesian Belief Networks and utilizes a set of analytical methods for the estimation of the risk model parameters.

Finally, a case study is presented, in which the risk framework developed here is applied to a maritime transportation system operating in the Gulf of Finland (GoF). The results obtained are compared to the historical data and available models, in which a RoPax was involved in a collision, and good agreement with the available records is found.

KEYWORDS

Maritime transportation; RoPax safety; Risk analysis; Bayesian Belief Networks; *F–N*diagram; Ship–ship collision

1. INTRODUCTION

Maritime traffic poses various risks in terms of fatalities, environmental pollution, and loss of property. In particular, accidents where ships carrying passengers are involved may pose a high risk with respect to human casualties. Therefore a number of studies on improvements to ship safety have been made; see for example [1], [2], [3], [4], [5] and [6]. One of the outcomes of these studies is the concept of risk-based design (RBD) for ships carrying passengers [7] and [8] where the major criterion for RBD is the ability of a ship to survive in damage conditions; see [9] and [10]. The above-mentioned studies address ship design; however, less attention has been paid to the risk-based design of ship operations. Although a general framework for this purpose is provided by the International Maritime Organisation – see [11] – few researchers have approached this topic in a holistic manner; see [12], [13], [14] and [15]. These models rely on accident statistics, and therefore the influence of factors contributing to the risk can hardly be measured. Moreover, most of the models utilise the concept of a fault tree (FT) or event tree (ET) following Boolean logic [16], [17] and [18], which in some cases may not fully reflect reality, as the events being analysed may take more than just two states. Furthermore FT and ET allow one-way inference, which in turn may limit their applicability in the field of systematic risk mitigation and management. The above limitations have been recognised in the field of Probabilistic Risk Assessment (PRA) in complex socio-technical systems, where alternative, hybrid approaches have been proposed, utilising FT, ET and Bayesian Belief Networks, see for example [19], [20], [21], [22], [23] and [24]. However, for the domain discussed in this paper, such solutions do not exist.

Therefore, it is desirable to develop a framework that evaluates the risk to ships at the design and operation stage in a proactive and systematic way. This would allow an insight to be gained into the process of risk evolution, as well as defining the most significant and sensitive variables that contribute the most to the risk in order to mitigate it in an optimal way, see [25] and [26].

Hence, this paper introduces a systematic, transferable and proactive framework determining the risk resulting from an open sea collision involving a RoPax. When it comes to describing the evolution of the accident the framework attempts to capture the causality, which makes the framework systematic. Its modular nature allows continuous improvement and adaptation to various locations and conditions, thus making its transferable. The framework is developed using Bayesian Belief Networks (BBNs), as recognised tools for knowledge representation and efficient two-way reasoning under uncertainty, which in turn makes the framework proactive; see [27], [28] and [29]. Moreover, BBNs allow reasoning in both directions, pointing out the most vulnerable nodes and the most effective ways of improving the outcome of the model. Thus the back-propagation of the probabilities can be utilised in the recommendation phase of risk assessment, see for example [30]. However, this is beyond the scope of this paper. All the above, along with the quantification of the effect of changes in the assumptions on the outcome of the framework – see [31] – make the results which are obtained more credible. Ultimately, the framework communicates to the end-user the level of available background knowledge about the system that is analysed, and how it is distributed across the framework.

Moreover, BBNs allow the adaptation of a formal risk definition following the well-founded idea of triplet given by Kaplan in [32]. Triplet attempts to answer the following questions: what can go wrong in the system?; how likely is it that it will go wrong ?; and what are the consequences if the assumed scenario occurs? This idea has been widely used for risk assessment and management in the maritime, see for example [33] and [34], which can provide additional justification for its use for the purpose of this study. All the relevant components of risk triplet are discussed here, and a case study is presented that addresses the maritime traffic system (MTS) operating in the Gulf of Finland (GoF) during the ice-free season, considering a specific RoPax as the struck ship.

The paper is structured into six main sections including the introduction. Section 2explains the use of BBNs to describe the risk in the MTS, with the adopted risk perspective. Section 3 defines the risk framework, and Section 4 describes the elements of the framework. In Section 5, the results obtained are shown, discussed and compared with the available data. Section 6 concludes the paper, focusing on the main findings and limitations of the risk framework.

2. BBNS AS REFLECTION OF RISK PERSPECTIVE

A formal, and well-established definition of risk in decision analysis is "*a condition under which it is possible both to define a comprehensive set of all possible outcomes and to resolve a discrete set of probabilities across this array of outcomes*", see [35]. To define a set of outcomes, knowledge and proper understanding of the system or phenomena being analysed is a prerequisite. This in turn enables scenarios leading to the outcome of interest and their probabilities to be defined. The framework aiming at risk analysis, which is presented in this paper, is developed by means of BBNs. These probabilistic tools allow reflection of the available knowledge on the process being analysed and its understanding in a comprehensive way. First, the aleatory uncertainties inherent to model variables are addressed, by describing variables using distributions obtained in the course of numerical analyses. Second, epistemic uncertainties related to model structure are analysed by performing alternative hypotheses testing. For this purpose, a set of scenarios is developed, with the constant set of variables, but different, plausible hypotheses governing the links between variables. This is performed for the elements of the model, which, for various reasons, can not be quantified with desired accuracy at the time of analysis. Beside quantification of the uncertainties the framework allows distributing them across the model. By doing this, the crucial areas of the model are pointed, where the limited background knowledge needs to be improved, as it has significant effect on an outcome.

Finally the framework communicates its output in a form of diagram, which presents the cumulative distribution of likelihood for the occurrence of number of fatalities given the scenarios.

By adopting this framework, and BBNs as tools for probabilistic modeling, it is possible to apply modified risk perspective of Kaplan [32], which for this paper reads as follows:

$$R = \{(S, L, C) | BK\} \tag{1}$$

where S stands for a set of scenarios which comprises the same chain of events, described by the same explanatory variables but the variables and their relations can be described by adopting different assumptions (alternative hypotheses). The latter depends on our background knowledge of the process being analyzed – BK, see for example [36]. L is a set of likelihoods corresponding to the set of consequences C, for a given set of scenarios (S) and given combination of anticipated assumptions governing the model parameters.

The above equation makes S, L, C conditional upon BK, which is in line with the formal definition of risk provided in the first paragraph in this section, adapted from [35]. This implies the necessity of quantifying the effect that BK, which for some variable may be limited, has on R.

BBNs as probabilistic tools for reasoning under uncertainty are capable of reflecting the analyzed scenarios along with assigning the associated probabilities. This in turn leads to the quantification of the consequences.

Moreover, the BBNs can effectively quantify the effect of limited knowledge and imperfect understanding of the analyzed system on the outcome of the framework.

This section discusses the main features of BBNs, stemming from the formal definition of risk. It also claims the BBNs to be suitable and proactive tools for risk evaluation in an MTS.

2.1. Background knowledge – BK

A clear representation of *BK* that is available about the given system is relevant for any model which is intended for practical use. This is especially important in the case of risk modelling in MTS, where *BK* about the system being analysed is limited and unequally distributed across the system. This, in turn, may introduce varying uncertainties depending on the elements of the system. This may result in a situation in which certain areas of the modelled system may lack a sufficient level of *BK* to satisfy the adopted formal definition of risk. Therefore, it is desirable for a risk framework to communicate the level of *BK* in order to determine whether the risk results are informative and can be used for decision-making, or should be used with great caution. The latter may be the case if the uncertainties are greater than the margin between the estimated risk and the risk limit. This may lead to some early stage conclusions about the appropriate level of granularity of the problem being analysed or the quality of *BK*, or the information that is available.

BK can be reflected in a systematic way by constructing a risk framework using BBNs. As a probabilistic graphical model, BBNs allow reasoning under conditions of uncertainty, as well as in the presence of limited data; see [37], [38] and [39]. Additionally, BBNs can be tested to find the essential variables which have the greatest impact on the output of the framework, and to determine which of these are the most informative; see [40]. For this purpose, a sensitivity analysis and a value-of-information analysis are performed. Moreover, as a result of the efficient updating of the outcome of the BBNs, given new knowledge on a variable or set of variables, the effect of changes in the predefined relations between variables, called likelihood functions (*LFs*), can be quantified. In this paper, this is accomplished by performing a so-called influence analysis. This analysis is especially important in the case of *LFs* which are not based on solid foundations.

All of these analyses allow BBNs-based risk framework to represent the level of available*BK* about the domain in question in a transparent and systematic way.

2.2. Scenarios – *s*

A fundamental stage of any risk analysis, and one which affects all the following stages, is scenario identification. This includes proper description of the knowledge about an MTS and its behaviour in a certain situation (e.g. an accident befalling a ship). This means that a risk framework should be capable of reflecting the right variables in the right way, considering the associated uncertainty along with a clear definition of the initial assumptions. Moreover, it should be able to determine the effects of the uncertainties and assumptions on the outcome of the framework, see [25], [41], [42] and [43].

Most of the existing models adopted for risk assessment in maritime transportation are defined in a spatio-temporal, stochastic framework; for a review of the models, see for example [44], [45], [46], [47] and [48]. However, these models often disregard causal relationships between input variables (e.g. ship size, collision speed, collision angle, relative striking location, and weather) and output variables (e.g. the ship capsizing). These relations are hidden under single probabilities (e.g. the probability of flooding given a collision or the probability of a severe collision) or probability density functions (e.g. a PDF representing the extent of the damage caused by a collision). This way of representing data disregards the causality in the scenario, and therefore substantial elements of risk analysis are missed, i.e. the links among variables and their mutual relationship, see for example [42]. This ultimately increases the uncertainty of the model.

However, some of the above-mentioned shortcomings of the existing models can be addressed by applying BBNs to a risk-analysis framework. First, BBNs allow multi-scenario thinking, which not only focuses on an undesired end event (a collision) but also provides insight into the process of the evolution of an accident. Second, BBNs structure reflects the causality in the process being analysed allowing further knowledge-based decision-making [49] and [30]. Third, BBNs can efficiently handle the uncertainties about variables and the uncertainties about the relations among variables, and represent those in the outcome.

2.3. Likelihood – *L*

In the field of risk analysis in engineering systems, three methods of interpreting the likelihood are usually followed: the relative frequency, subjective probability and a mixture of these called the probability of frequency; for discussion on these see, for example, [32],[50] and [51]. The BBNs described in this paper combine numbers stemming from the first two concepts, whereby the decision of which to adopt was made with respect to the available knowledge. If the latter permits, the frequentist approach is used, and a repetitive experiment is conducted; otherwise the probabilities are derived through the elicitation of knowledge from experts. The probabilities, as mathematical concepts, follow certain axioms, which in some real-life cases may not hold true; for discussion on the axioms and validity of the given approaches, see, for example, [51] and [52].

The numbers derived from various sources are combined with the use of BBNs, which encode the probability density function governing a set of random variables by determining a set of conditional probability functions (*CPFs*). Each variable is annotated with a *CPF* , which represents the probability of the variable given the values of its parents in the graph ($P(X|pa(X)) \in P$). The *CPF* describes all the conditional probabilities for all the possible combinations of the states of the parent nodes. If a node does not have parents, its *CPF* reduces to an unconditional probability function, also referred to as a prior probability of that variable.

From a mathematical viewpoint, classical BBNs are a pair $N=\{G,P\}$, where $G=(V,E)$ is a directed acyclic graph (*DAG*) with its nodes (V) and edges (E), while P is a set of probability distributions of V. Therefore, BBNs representing a set of variables and their dependencies consist of two parts, namely a quantitative (P) and a qualitative (G). Therefore, a network $N=\{G,P\}$ is an efficient representation of a joint probability distribution $P(V)$ over V, given the structure of G following the formula; see also[27] and [28]:

$$P(V) = \prod_{X \in V} P(X \mid pa(X)) \qquad (2)$$

In the framework described here, the *CPFs* were obtained through simulation, a literature study, natural laws and expert opinions. The *CPFs* are relevant elements of the framework; first, they govern the flow of knowledge through the framework, and second, they constitute a link between the qualitative and quantitative parts of the framework.

3. RISK FRAMEWORK DEFINITION

This section describes the five-step procedure according to [53], defining the risk framework as follows:

1 defining what to model;
2 defining the variables;
3 developing the qualitative part of the framework;
4 developing the quantitative part of the framework;
5 validating the framework.

3.1. Defining what to model

The aim of the proposed framework it to estimate the risk in MTS, focusing on selected accidental scenarios that, ultimately, lead to the loss of a struck RoPax ship. These scenarios are (i) the inner hull of the RoPax that is struck is breached and consequent flooding is experienced; this can result further in the loss of the ship; (ii) the RoPax that is struck has no significant hull damage; however, the ship is disabled and drifts, thus experiencing significant rolling as a result of wave and wind action, which can result further in the ship capsizing. The loss of the RoPax is expected if two consecutive limits are exceeded, namely crashworthiness and stability.

Subsequently the corresponding probabilities of the limits being exceeded given the traffic and environmental conditions are evaluated on the basis of the model presented here. For this purposes the following general factors are taken into consideration: the composition of the maritime traffic in the sea area being analysed, the collision dynamics, hydrodynamics of the ship and her loading conditions.

Ultimately, the cumulative number of fatalities (*N*) resulting from an accident is modelled utilising the concept of the rate of fatalities. This rate is determined taking into account time for evacuating a ship and time for a ship to capsize. The number of passengers on board is modelled utilising available data from RoPax operators from the Gulf of Finland. All these, along with the associated probabilities (*P*) for a given number of fatalities, are finally depicted in a F−N diagram, which can be considered as a risk picture.

3.2. Defining the variables

The framework presented here attempts to reflect the causality in the process of open-sea collision that is being analysed by defining the relevant variables and constructing logical relations between them. Thus the framework consists of four major parts, covering the following areas: (i) collision-relevant parameters; (ii) capsizing-relevant parameters; (iii) the response to an accident; (iv) quantification of the consequences. These are described further in Section 4, and are depicted in Fig. 1. The collision-relevant parameters are obtained from a maritime traffic simulator, which utilizes AIS data and accident statistics for the Gulf of Finland, see Section 4.1. Ship capsizing is conditional upon various events, of which the most relevant are (i) the collision speed and angle for the given ship mass ratios, leading to the rupture of the inner hull of a struck RoPax conditional upon a collision; (ii) the extent of damage leading to the significant ingress of water, conditional upon the inner hull being ruptured; (iii) the hydro-meteorological conditions contributing to the ship capsizing given the significant ingress of water; (iv) the maximum roll angle at which a disabled, intact RoPax capsizes. All these are obtained with the use of finite element simulations, a six-degrees-of-freedom ship motion model, and the available literature.

The response to an accident means the evacuation time or time needed for rescue tugs to arrive to the accident place. The former is modelled adopting the IMO requirements. The latter with the use of maritime traffic simulator and available data about the location of the rescue tugs in the area.

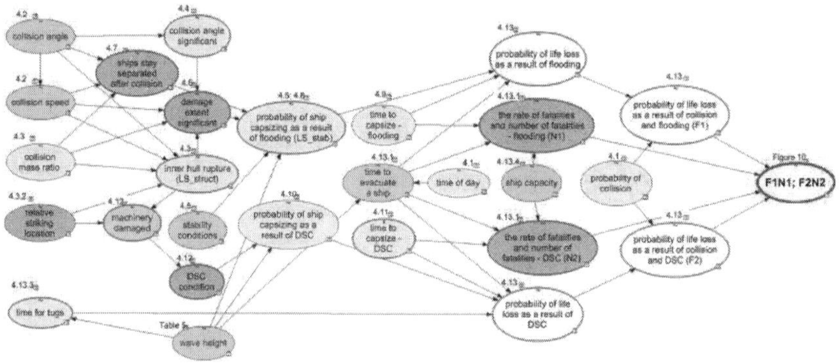

Figure 1. The qualitative description of the risk framework introduced here. At each variable the reference to a section which describes a given variable is provided. (For interpretation of the references to color in this figure caption, the reader is referred to the web version of this paper.)

The consequences of an accidents are modelled with the use of a concept of the rate of fatalities, as described in Section 4.13.1.

3.3. Developing the qualitative part of the framework

In this step, the graphical structure of the network is created. As available accident databases are scarce with respect to the consequences for a RoPax ship that is struck by another ship, another source of information must be found. We decided to utilise the qualitative part of the existing ET-based risk framework for RoPax, see [3] and [54], and confront it with expert knowledge about the domain.

As the domain under study is wide and multidisciplinary, we divided it into the following sections: (i) ship operations, including the stability of the ship, (ii) the structure of the ship, and (iii) accident response.

Expert knowledge about the domain was elicited through a brainstorming session as well as individual meetings. During the session, the initial structure of the model was presented to the experts for assessment, and was then modified according to their suggestions. The group of experts consisted of 15 researchers and practitioners in the fields of ship design, ship operation and rescue services. Once the structure of the model had been defined, the quantitative part of the model was determined.

3.4. Developing the quantitative part of the framework

The number of probabilities required for a BBN depends on the structure of the network and the number of variables and their states, and it grows exponentially. To reduce the number of probabilities that need to be determined to evaluate the framework, the parametric probability distributions (PPDs) for the variables were used. These provide simple computation rules for obtaining the required probabilities; see [29]. All the PPDs applied in the model are described in detail in Section 4.

A complete list of the model variables and their PPDs is presented in Appendix A in Table A1, which also lists the data sources for the variables.

3.5. Validating the framework

At this stage, the risk framework is validated by performing the following; see also [40]: sensitivity analysis of the framework, value-of-information analysis, influence analysis and a comparison of the results obtained with the available data. This stage is important in the context of the reliability and validity of the framework, and knowledge distribution and uncertainty analysis in the framework and its output; see [55] and [43].

The lack of *BK* about the analyzed system leads to uncertainty in the model parameters and affects the hypotheses supporting the model structure. There are numerous ways to address and express the model uncertainty, see for example [56]. The risk framework introduced here allows for the effect of uncertainties related to *BK* being evaluated in two-fold. First, by analyzing aleatory uncertainties of the variables. For this, the relevant variables are considered as distributions and the analysis is performed for a range of parameters that these distributions can take. Second, epistemic uncertainties related to model structure are analysed by performing alternative hypotheses testing. For this purpose, a set of models (BBNs) is developed, with the constant set of variables, but different, plausible hypotheses governing the links between variables. In addition to quantifying the uncertainties, the framework allows for their distribution across the model. By doing this, the crucial areas of the model are identified. In these areas the limited *BK*must be improved, as it has significant effect on the outcome.

If the framework is used for analysing the risk in a specific maritime transportation system, the sensitive variables should be evaluated locally in order to reflect the actual conditions. However, if this is not possible, the framework allows the quantification of the effect of underlying assumptions on the outcome of the framework. The remaining less sensitive variables of the framework can be more generic, as changes to them do not affect the outcome significantly.

4. RISK FRAMEWORK AGGREGATION

This section describes the methods adopted to determine the risk framework, meaning the variables and the relations among them. It also presents the results of the performed case study, which focused on the MTS operating in the Gulf of Finland (GoF). Although the results are valid for the maritime traffic composition and hydro-meteorological conditions prevailing in the GoF, the methods applied are generic and the modularity of the framework makes the approach presented here fully transferable. The framework is encoded with BBNs, developed by means of an available software package called GeNIe; see [57] and [58]. The qualitative description of the framework is given in Fig. 1, where each variable is annotated with the reference to a section which describes this variable. Moreover, three colours are used to make a distinction between variables which are obtained from the numerical simulations (blue), taken from the literature (yellow), based on certain assumptions (grey) or purely conditional on their parents (without filling).

The framework captures the accidental scenario, where a RoPax is struck by another ship in the open-sea. The collision is described by collision angle, collision speed and masses of colliding ships and relative striking location along a hull. Potential locations of an accident and collision related parameters are obtained from a maritime traffic simulator. To determine whether the inner hull of a RoPax is ruptured following the collision, the critical speed is calculated with the use of numerical model and then it is compared with the actual collision speed. If a collision speed is higher than critical speed, a hull rupture is expected. Then the framework estimates whether the damage can cause catastrofic flooding. If this is the case, the model estimates the probability for a ship to capsize along with time to capsize. A RoPax may also capsize even if she is intact, if she is exposed to wave action. Therefore, the probability for a RoPax which is intact but disabled due to post-collision damage to her propulsion is calculated with the use of ship dynamic model. For the probability and time to capsize, a set of stability conditions is considered along with the wave characteristics typical for the GoF. The time needed to evacuate the RoPax is modelled with the use of IMO recommendations, which leads to quantification of the rate of fatalities, given collision and flooding. This, together with the number of people on board, estimates the cumulative probability (F) of a given number of fatalities (N). The framework delivers its output in a form of F−N diagram, which is recognized way of risk visualization.

This section provides description of all variables included in the framework.

4.1. Collision probability

In the case study presented here, the probability of a collision between two ships in the open sea in which the RoPax is struck by another ship is estimated by means of the dynamic maritime traffic simulator (DMTS), developed by Goerlandt and Kujala in [45]. Additionally the accident statistics compiled for the GoF are utilized. The input to the DMTS is taken from the Automatic Identification System (AIS), augmented with harbour statistics concerning the cargo types that are traded. The model determines the annual frequency of a collision between two ships assuming a RoPax being struck for the whole GoF, as if ice-free conditions were year-round. The annual frequency of such an accident equals 0.1. This means that a collision in which a RoPax is involved would happen every 10 years, if there were ice-free conditions year-round. However, this is not the case for the GoF, therefore more realistic assumptions are made that the ice is present every year and remains there for 3 months. This means that the annual frequency for an open-sea collision in which a RoPax is struck equals 0.07, and the reoccurrence period for such an accident is 14 years.

However, the available accident statistics which are compiled for the GoF reveal that the annual frequency of the collisions between two ships at sea, regardless of ship type and ice conditions is 0.2; see [59]. Then, attributing equal chances of being struck and striking to a ship involved in a collision, the annual frequency for a ship being struck, regardless of her type, given a collision is 0.1. Assuming that the ratio of RoPax ships to the number of other ships navigating in the GoF is 1:10, the annual frequency of a RoPax being struck in a collision yields 0.01. Assuming, period of 3 months, during which the GoF is frozen, the frequency of an open-sea collision, where a RoPax is struck is 0.0075.

Thus, two numbers for the annual frequency of an accident in the open-sea in which a RoPax is struck are obtained, 0.07 from the DMTS, and 0.0075 from the accident statistics. However, both values are burdened with some amount of uncertainty, due to assumptions in the DMTS or simplifications in reasoning from the accident statistics. Therefore it is assumed that the "true frequency" might fall between these two numbers, and they are considered as limits for a uniform distribution estimating the probability of an open-sea collision, where a RoPax is a struck ship; see Eq. (3).

The most likely locations of such an accident are depicted in Fig. 2, and the simplification is made about lack of correlation between time of day and collision:

$$P_{coll} = \begin{cases} \dfrac{1}{0.07-0.007}, & 0.007 \le x \le 0.07, \\ 0, & x < 0.007 \text{ or } x > 0.07 \end{cases} \qquad (3)$$

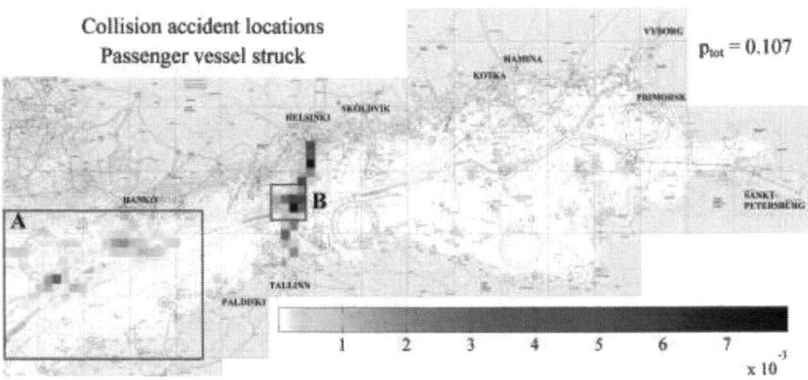

Figure 2. The modelled locations of an accident in which a RoPax is the struck ship.

4.2. Collision parameters

Another item of information derived from the DMTS concerns maritime traffic data in terms of the composition of the traffic, ship types, ship sizes, collision angles, collision speed and the time of day of a potential collision. These are essential input for the risk framework presented here, as they describe in detail the tempo-spatial layout of traffic. The DMTS generates a trajectory for each single vessel sailing in the area, called a traffic event, and assigns a number of parameters to this event, as illustrated in Fig. 3.

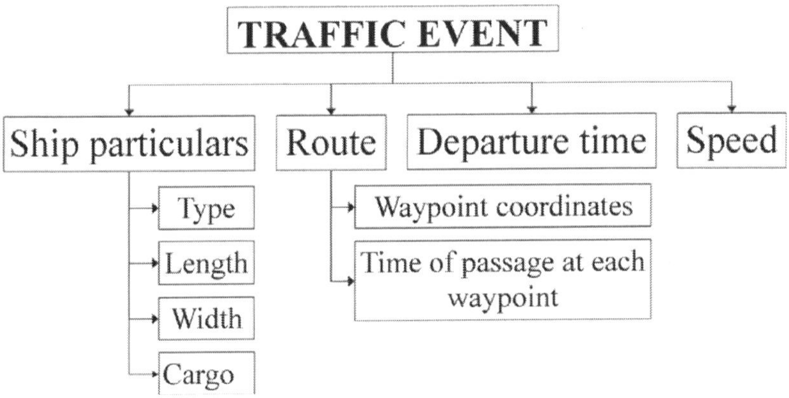

Figure 3. Data generated for each simulated vessel, a traffic event [45].

The traffic events modelled by the DMTS are based on data which refer to the normal operation of ships, ultimately resulting in safe navigation. Therefore, the modelled motion parameters of ships on a collision course (their speeds and courses) do not account for the changes caused by evasive manoeuvres intended to forestall a collision. To fill this gap, a two-step procedure is applied to determine the collision speed. First, the initial value of this parameter is obtained from the DMTS, and then it is considered as the input value for the statistical models, thus arriving at the actual collision speed. There are several different statistical models for estimating the collision speed and collision angle; see [42]. However, only one, proposed by Lützen in [60], takes into account the changes in the initial parameters resulting from evasive action taken by the colliding ships. Therefore, this concept is applied here with the following assumptions:

1 the velocity of a striking ship *A* follows a uniform distribution for velocities between zero and 75% of her initial speed, then the probability decreases triangularly to zero at her initial speed;
2 the velocity of a struck ship *B* is approximated by a triangular distribution with the most likely value equal to zero and a maximum value equal to her initial speed;
3 the initial speed values of *A* and *B* are obtained from the DMTS;
4 the collision angle, defined as the difference in the headings of two colliding ships, is uniformly distributed between 10° and 170°.

Then, applying the four-step random sampling Monte Carlo procedure, the distribution of the actual collision speed is estimated as follows:

1 sample the initial speed of a striking ship obtained from the MDTS, then use it as an input to determine the appropriate uniform-triangular distribution; subsequently sample the speed from this distribution randomly, and store it as *VA*;
2 sample the initial speed of a struck ship obtained from the MDTS and use it as an input to the triangular distribution; subsequently sample the speed from this distribution randomly, and store it as *VB*;
3 randomly sample the collision angle α from the uniform distribution;
4 knowing the *VA*, *VB* and α, calculate the relative speed at which ship *A* hits ship *B*, consider it as the collision speed, [61] and store it as $V_{(A,B)}$; see Fig. 4.

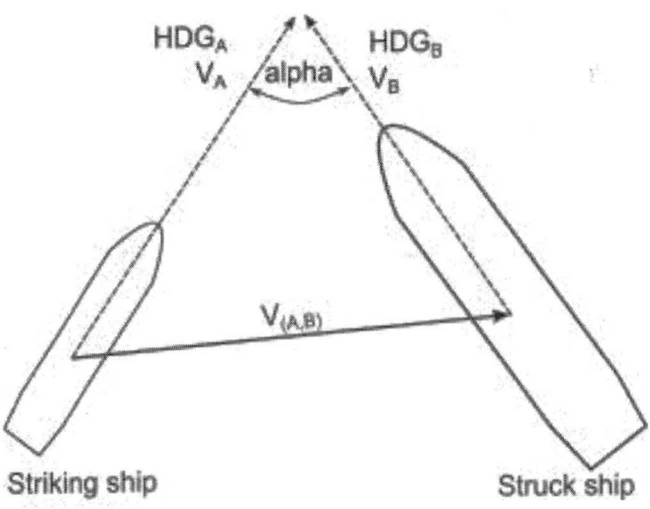

Figure 4. Definition of collision speed – $V_{(A,B)}$.

For each collision encounter obtained from the DMTS, the above procedure is run repeatedly, and a set of collision speeds given a certain collision angle is arrived at. Then, all the collision speed values that have been obtained are ordered according to the collision angle and are divided into three sub-sets: 10–45°, 45–135°, and 135–170°. Subsequently, the sub-sets are described with the use of normal distributions, which are the best-fits, see Fig. 5. Finally, these distributions are embedded into the BBNs presented here, as demonstrated in Table 1.

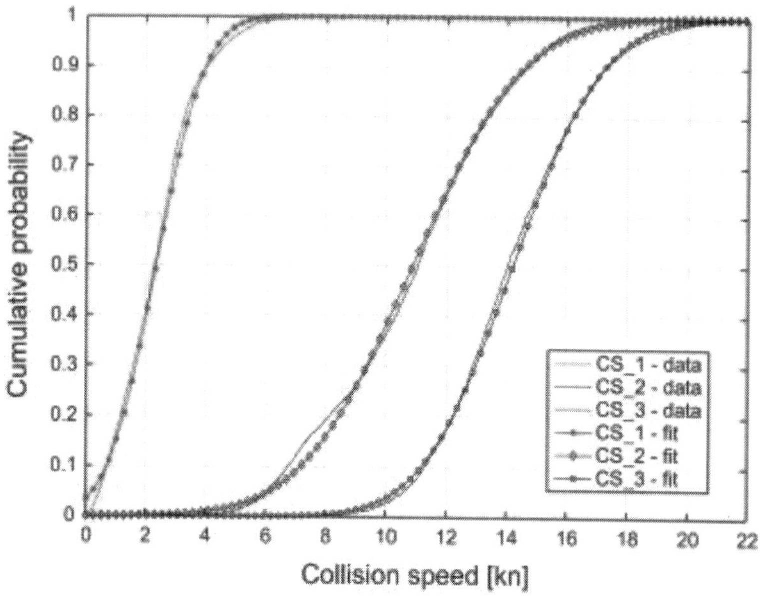

Figure 5. The CDFs of variable *collision speed* – labelled as *CS* in the graph – conditional upon the *collision angle* adopted in the BBNs developed here.

Table 1. The continuous probability distributions for variable *collision speed*, given the *collision angle*.

Collision angle (deg)	Collision speed (kn)
10–45	$CS-1=N(2.35, 1.64)$
45–135	$CS-2=N(10.86, 8.26)$
135–170	$CS-3=N(14.2, 5.53)$

4.3. Rupture of the inner hull of the RoPax

The value of the actual collision speed is one of the inputs to a function determining whether the inner hull of a RoPax is ruptured (*ihr*); see Eq. (4). The other input is the structural capacity of the ship being analysed, described by the function called *limiting speed – Vrupture*, which is a speed leading to inner hull rupture. This parameter is evaluated with the use of a numerical model, which is described in this section.

The relation between the collision speed – $V_{(A,B)}$ – and the limiting speed – *Vrupture* – is given by the following Heaviside function:

$$ihr = \begin{cases} 0, & V_{(A,B)} < V_{rupture} \\ 1, & V_{(A,B)} \geq V_{rupture} \end{cases} \tag{4}$$

The *Vrupture* quantity is a function of the sizes of the colliding ships, the striking angle and the relative striking location along the hull of a struck RoPax. This quantity is determined using the concept of collision energy, which is evaluated for the reference RoPax; see Table 2. In the case study presented here, the following sizes of the striking ship are considered with respect to the struck RoPax: a similar size (mass ratio 1.0), a ship that is 25% smaller (mass ratio 1.33), a ship that is 25% larger (mass ratio 0.8) and a ship that is 70% larger (mass ratio 0.6). Therefore the mass ratios that are analysed cover almost 80% of maritime traffic in the GoF. The remaining share belongs mostly to ratios higher than 1.3 and lower than 0.6, which are not taken into account in the numerical analysis. Ratios higher than 1.3 can be assumed to be less critical concerning a hull rupture for the usual blunt bow shapes. However, ratios lower than 0.6 shall not be neglected, thus despite the fact that they are excluded from the numerical analysis, they exist in the risk model. The conservative assumption is made that in any case where the ship masses ratio is lower than 0.6, the inner hull of a RoPax is breached, if the collision speed is higher than 80% of the limiting speed for the ratio 0.6.

$$ihr(ratio < 0.6) = \begin{cases} 0, & V_{(A,B)} < 0.8V_{rupture}\ (ratio = 0.6) \\ 1, & V_{(A,B)} \geq 0.8V_{rupture}\ (ratio = 0.6) \end{cases} \quad (5)$$

Table 2. The characteristics of the RoPax vessel that was analysed.

Length	188.3 m
Breadth	28.7 m
Average draught	6.0 m
Displacement	19 610.0*t*

The available energy for structural deformations is obtained according to the calculation model introduced by Tabri in [62]. His model estimates the dynamics of a ship collision and the share of energy available for ship motions and structural deformations. As a result of the combination of this dynamic simulation procedure and the non-linear finite-element method, a good estimation of structural damage in various collision scenarios with oblique angles and varying eccentricities of the contact point can be achieved, see also [63]. The simulated collision scenarios, defined by the striking angles and relative striking locations along the hull of the RoPax, are depicted in Fig. 6.

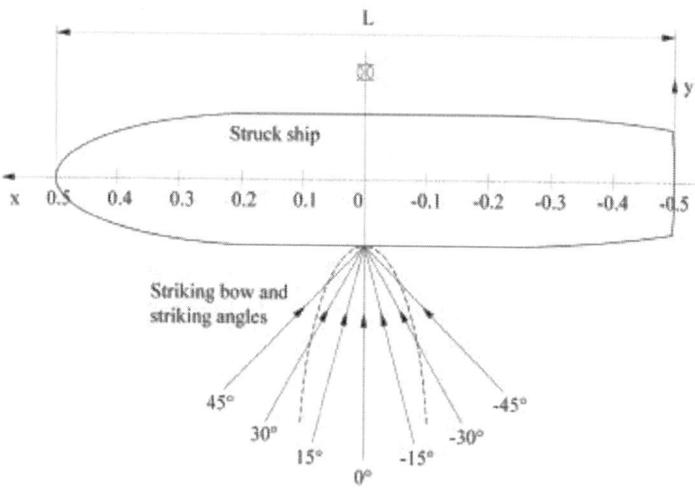

Figure 6. Relative striking locations and striking angles analysed.

For the purpose of collision simulations, the LS-DYNA solver version 971 is used, and the ANSYS parametric design language is used to build the finite-element model of the reference RoPax vessel. A three-dimensional model is built between two transverse bulkheads of the midship section, spaced 26.25 m apart – see Fig. 7 – and not accounting for differences in the ship structures present along the length of the ship. Moreover, the translational degrees of freedom are restricted in the plane of the bulkhead locations, whereas the remaining edges are free. The structure is modelled using four-noded, quadrilateral Belytschko–Lin–Tsay shell elements with five integration points through their thickness. The characteristic element length in the contact region is 50 mm in order to account for non-linear structural deformations, such as buckling and folding. The element length-dependent material relation and failure criterion according to [64] are utilised for the simulations. Crashworthiness simulations employing this material model have been found to be sufficiently accurate compared to large-scale experiments; see [65]. Standard LS-DYNA hourglass control and automatic single surface contact (with a friction coefficient of 0.3) are used for the simulations. Moreover, the collision simulations are displacement-controlled.

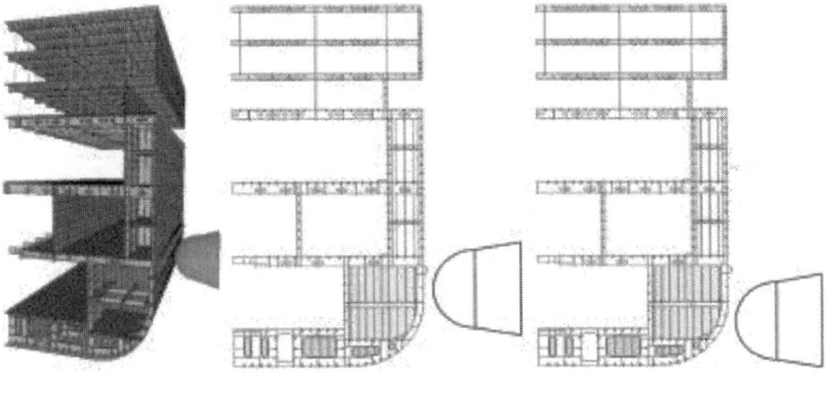

Figure 7. FEM model and vertical striking locations.

The rigid bow is moved into the side structure of the ship in a quasi-static fashion. Hence, this approach results in the maximum absorption of energy by the side structure alone, which is needed for a comparison and can be considered conservative and therefore suitable for fast prediction. Moreover, two draughts are considered for the struck ship, which are a function of the maximum variations in draught to be expected for this kind of vessels. Therefore the maximum striking location is positioned close to the first deck, and the second striking location is located closer to the tank top. Then it as assumed that both locations are equally probable and thus take the average response.

As a result, the relative energy available for structural deformations as a function of the longitudinal striking location is obtained for a mass ratio of 1.0; see Fig. 8. For mass ratios of 1.33 and 0.6, these curves are scaled with the factors 0.84 and 1.13, respectively, to account for the change in dynamic behaviour. Therefore, the values of *Vrupture* for a given mass ratio, striking angle and striking location along the hull of the struck ship, causing a breach of her inner hull, are evaluated; see Table 3.

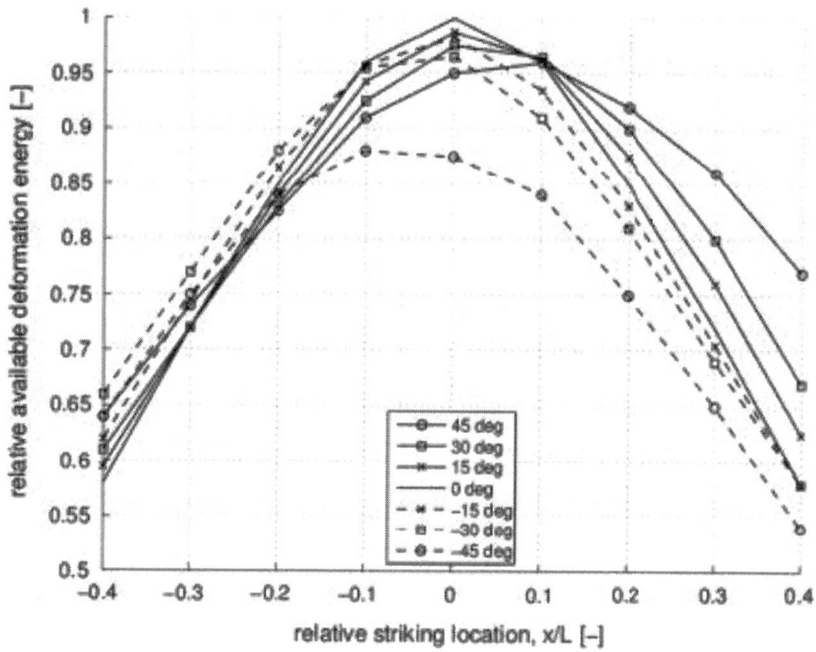

Figure 8. Relative available deformation energy versus relative striking location and striking angle.

Table 3. The values of the structural capacity for a RoPax, expressed by limiting speed – in knots – as a function of the relative striking location and striking angle, mass ratio 1.0.

Relative striking location	Striking angle (deg)						
	45	30	15	0	−15	−30	−45
−0.4	8.20	8.40	8.50	8.61	8.33	8.07	8.20
−0.3	7.63	7.73	7.73	7.73	7.60	7.48	7.57
−0.2	7.22	7.20	7.16	7.12	7.05	6.99	7.16
−0.1	6.88	6.82	6.76	6.70	6.70	6.71	6.99
0.0	6.73	6.64	6.60	6.56	6.62	6.68	7.01
0.1	6.70	6.68	6.69	6.70	6.78	6.88	7.16
0.2	6.84	6.91	7.01	7.12	7.20	7.29	7.57
0.3	7.07	7.73	7.52	7.73	7.81	7.90	8.14
0.4	7.48	8.01	8.29	8.61	8.61	8.61	8.92

Finally, the probability of the inner hull rupture given a collision – *Pihr* – is obtained with the use of the framework introduced here. For the case study analysed here, it yields $Pihr=0.61$.

4.3.1. Masses of colliding ships

In the case study presented here, the masses of colliding ships are obtained from the DMTS, and then modelled with the use of a log-normal distribution with the following parameters: μ=0.25 and σ=0.70.

4.3.2. Relative striking location

Relative striking location is modelled with the use of a uniform distribution. This means that any location along the hull of a RoPax is equally probable for being hit by the striking ship. Thus, the limits for this distribution are −0.5 (the most aft part of the RoPax) and 0.5 (the most forward part of the RoPax); see Fig. 6.

4.4. Collision angle significant

It is assumed that the collision angles falling in a range between 45° and 135° may lead to a rupture of the inner hull of a RoPax, and thus they are considered relevant, see also[66]. Therefore the collision angle from this range is referred to as *α significant* (*αsign*). This assumption corresponds to the range of striking angles adopted in the FEM-based RoPax crashworthiness experiment, which is described in the previous section.

Nevertheless, the effect of a change to this assumption on the outcome of the framework and relevant intermediate variables is determined at the validation phase of the framework.

4.5. Capsizing of a damaged ship as a result of flooding

As a result of a ship–ship collision in which the collision speed exceeds the limiting speed for a given RoPax in given scenario (size of colliding ship, collision angles and location), the ingress of water can be expected. This, in turn, can bring the damaged ship being analysed towards the boundaries of her stability limit. In the risk framework presented here, it is developed utilising the concept of a "capsize band," determined on the basis of numerical simulations that consider characteristics of the ship, her loading conditions and environmental properties as explanatory variables. The band is determined for a given ship, as a function of wave height (*H*) and ship stability (*stab*); it also takes dynamic time-variant flooding characteristics into consideration, see [67] and [68]. Within the band a transition between two states, namely "safe" and "unsafe," takes place, according to the logic presented in Eq. (6). The band begins at a wave height that does not cause the ship to capsize ($Hcapsize_{=0}$) and ends at the wave height where the loss of the ship is always expected ($Hcapsize_{=1}$). The capsize boundaries are symmetrical around the value of the critical wave height (*Hcritical*), which corresponds to the probability of capsizing equal to 0.5. The band is estimated with the use of a sigmoid function (*S*). This logic is captured by the following function:

$$P_{capsize}(stab) = \begin{cases} 0, & H < H_{capsize=0}(stab) \\ S(H_{critical}), & H_{capsize=0}(stab) \leq H \leq H_{capsize=1}(stab) \\ 1, & H > H_{capsize=1}(stab) \end{cases} \tag{6}$$

For a detailed description of the concept the reader is referred to [6] and [69].

The framework presented here assumes that the flooding contributing to the loss of the ship occurs if the wave is higher than a critical height, and at least the main car deck and two compartments beneath are flooded, which corresponds to the accident scenario adopted in the previous work of Papanikolaou et al. presented in [6]. For the purpose of this study, wave data for the sea area that is analysed is shown in Table 5. Four equally likely stability conditions of a damaged RoPax are assumed, si, where i =1–4. They correspond to the following $H_{critical}$ [m] = [2.0, 2.5, 3.5, 5.5], with appropriate bandwidths around each *Hcritical*, see Table 4. For a detailed description of the ship conditions and method, the reader is referred to [6].

Table 4. The capsize bands and likelihood functions applied in the model.

$H_{critical}$ (m)	Likelihood functions ($P_{capsize}$) and corresponding capsize bands		
	$P_{capsize} = 0$ (m)	$P_{capsize} = S(0,1)$ (m)	$P_{capsize} = 1$ (m)
2.0	0–1.3	1.3–2.8	> 2.8
2.5	0–1.5	1.5–3.4	> 3.4
3.5	0–2.0	2.0–4.8	> 4.8
5.5	0–3.5	3.5–7.2	> 7.2

Finally, the conditional probability of a RoPax exceeding her stability limits resulting in ship capsizing is obtained – *Pcapsize*. For the case study being analysed here, it yields $Pcapsize = 0.17$.

Table 5. The wave statistics for the Baltic Sea including the GoF, [70].

Wave height (m)	Probability of occurrence
0–1	0.345
1–2	0.390
2–3	0.185
3–4	0.062
4–5	0.015
5–6	0.003

4.6. Damage extent significant

The damage stability conditions analysed in this paper consider a RoPax experiencing the flooding of certain compartments, which are the main car deck and two of the compartments beneath. However, not every hull breach results in such severe effects, and, therefore, the conditional probability of the damage size (*des*) allowing critical flooding is estimated by considering the mass ratio of the ships colliding, the collision speed and collision angle, adopting the following function:

$$des = \begin{cases} 1 & \text{AND}(ihr) = 1,\ \alpha_{sign} = 1,\ V_{(A,B)} > 1.2V_{rupture} \\ 0 & \text{otherwise} \end{cases} \quad (7)$$

Here, the collision speed leading to significant damage is taken as 120% of the structural capacity for a RoPax, defined by the limiting speed, introduced in Section 4.3. The collision angle contributing to significant damage is called *collision angle significant*(α_{sign}), and it is introduced in Section 4.4.

In the framework presented here, the conditional probability of such a critical accident scenario is evaluated by a node called *damage extent significant* (*des*), which, for a given case study, yields *Pdes*=0.15. As this variable is quantified on the basis of an assumption, the analysis of the effect of a change to the assumption on the response of the framework is conducted at the validation stage.

4.7. Ships stay separated after collision

A RoPax suffering significant damage resulting from a collision may experience the rapid ingress of water if the opening caused by damage is exposed to high seas. This may occur if two ships that collided remain separated after the collision, instead of the striking ship having her forward part stuck in the side of the struck ship. The probability of these two ships being apart is governed by a function binding the following variables: *collision speed*, *collision angle significant*, and *collision mass ratio* (*cmr*), through the following Heaviside function:

$$sss = \begin{cases} 1 & \text{AND}(V_{(A,B)} > 1.2V_{rupture},\ cmr < 1,\ \alpha_{sign} = 1) \\ 0 & \text{otherwise} \end{cases} \quad (8)$$

The probability of such a situation for the case study presented here equals$P_{sss}=0.94$. As this variable is based on assumptions, an analysis of the effect of changes in the assumptions on the results of the framework is carried out.

4.8. Capsize resulting from flooding

The framework developed here assumes that a RoPax will capsize if the collision speed is higher than the limiting speed for the given ship and given collision scenario, the damage is significant, the two ships that have collided are separate after the collision and the stability limit is exceeded for a given stability conditions. Otherwise, the ship is not expected to capsize. The probability of this event is evaluated by the node called *ship capsizing as a resulting of flooding*, using the following formula:

$$P_{C-flooding} = \begin{cases} P_{capsize} & \text{AND}(des = 1,\ sss = 1) \\ 0 & \text{otherwise} \end{cases} \quad (9)$$

where *Pcapsize* is determined through Eq. (6). The probability of a RoPax capsizing as a result of flooding for the case study presented here yields $P_{C\text{-}flooding}=0.018$.

4.9. Time to capsize because of flooding

The time to capsize (*TTC*) is a relevant factor when it comes to the evaluation of the success of the evacuation of the ship once catastrophic flooding is experienced. This framework recognises this parameter, and for the case study presented, we use a probabilistic model based on the results of numerical simulations by Spanos and Papanikolaou in [10]. The model that is applied here reflects their findings, where the probability of the ship capsizing within 30 min of the damage event equals 0.8 and reaches 0.95 within 60 min. Therefore, the following function is adopted to determine*TTC*:

$$TTC\ [\mathrm{min}] = \exp(\lambda) \tag{10}$$

where λ=0.05, and the distribution is truncated at *TTC*=180 min, see [10].

4.10. The probability of a ship capsizing in dead ship condition

Another type of consequence arising from an open sea collision is a ship capsizing as a result of wave and wind action, where the ship is in dead ship condition (DSC). DSC means "*a condition in which the entire machinery installation, including the power supply,is out of operation and the auxiliary services for bringing the main propulsion into operation and for the restoration of the main power supply are not available*;" see also[71]. This phenomenon is dependent on the ship type, and thus the hull shape, and weather conditions. Thus, for the purpose of this case study, simulations are performed to obtain the probability of a RoPax capsizing as a result of DSC using the state-of-the-art, six-degrees-of-freedom (6-*DoF*) ship dynamics model; see [72]. The probability of the ship capsizing is assumed to be equal to the probability of a particular angle of roll being exceeded, in this case 60°. To calculate the probability of this roll angle being reached –$P_{C\text{-}DSC}$ – Monte Carlo simulations are applied:

$$P_{C-DSC} = \frac{N_{\phi_C}}{N_S} \tag{11}$$

where $N_{\phi C}$ means the number of simulations in which the roll angle that would lead to the ship capsizing was reached, and *NS* is the overall number of trials.

The 6-*DoF* model assumes that the overall ship response is a sum of linear and non-linear parts. Such a division is a result of the fact that the linear calculating methods are well known, and the hydro-mechanical radiation and diffraction forces are well presented by linear formulae. The main part of the first-order load is calculated with a linear approximation (added mass, damping coefficient), with the actual heading and placement with respect to the waves being considered, whereas the following are considered to be the non-linear parts: Froud–Krylov forces, restoring forces and non-linearity resulting from motion equations.

For the case study that is analysed, the probability of a RoPax capsizing, given the collision followed by DSC, yields $P_{C\text{-}DSC}=2.0\times10^{-4}$.

4.11. Time to capsize as a result of DSC

The ship dynamics model mentioned in the previous section simulates ship motions in the time domain. Thus, for each event associated with a ship capsizing, the time instants of this event are recorded. Monte-Carlo simulations are applied to evaluate the fractions of cases where a ship capsizes, and thus the distribution of the time instants taken a ship to capsize as a result of DSC is obtained. This is considered as an input variable to the risk framework. A log-normal distribution is used for modelling this parameter, where, μ=8.0 and σ=5.7, as follows:

$$TTC_{DSC}\ [\mathrm{min}] = \ln \mathcal{N}(\mu, \sigma^2). \tag{12}$$

4.12. Machinery damaged

The occurrence of DSC is governed by the unavailability of the ship's main propulsion or steering. The probability of the ship's propulsion and steering systems being damaged as a result of an accident is determined by the node called *machinery damaged*, given a RoPax being hit in the section housing the steering devices or main engine (*steer-me*) and having her inner hull ruptured (*ihr*). However an assumption is made about a 50% chance of the main propulsion or the steering gear failing as a result of the collision impact only, even if the inner hull is not ruptured. The probability for machinery damage is calculated with the following formula:

$$P_{md} = \begin{cases} 1 & \mathrm{AND}(ihr = 1, steer-me = 1) \\ 0.5 & \mathrm{AND}(ihr = 0, steer-me = 1) \\ 0 & \text{otherwise} \end{cases} \tag{13}$$

It is assumed that the length of the section accommodating the propulsion is 0.2 LOA, and thus *stree-me* is modelled as follows:

$$steer-me = \begin{cases} 1 & \mathrm{If}(Uniform(0, 10) > 8) \\ 0 & \text{otherwise} \end{cases} \tag{14}$$

The probability of the machinery being damaged as a result of a collision for the case study analysed here yields *Pmd*=0.16. At the stage of the validation of the framework, the effect of the changes in the above assumptions on the results of the framework is examined.

4.13. The probablity of loss of life and accident response

Two means of responding to an accidental collision to a RoPax are considered in this framework. First, a ship salvage operation with the use of tugs is considered in a case where a ship that has been in a collision experiences DSC but no flooding occurs. Second, an ordered evacuation of a ship takes place if there is serious flooding following the collision. If the response time (*resp*) is shorter than the hazard exposure time (*haz*), namely the time to capsize as a result of flooding or DSC following a collision, such a situation is considered as a success. Otherwise, the response is not effective and loss of life (*LL*) can be expected. The following Heaviside function is applied to determine this parameter:

$$LL = \begin{cases} 0, & haz/resp > 1 \\ 1, & haz/resp < 1 \end{cases} \tag{15}$$

For the case study presented here, the loss of life can be expected in 82 out of 100 cases in which a ship is flooded. In the cases in which a ship capsizes as a result of DSC, this ratio is even higher, and equals 98 out of 100.

To obtain the annual probability for loss of life resulting from a ship capsizing as a result of collision, the following conditional functions are adopted:

$$P_{LL|\,capsizing|\,collision} = P_{LL} P_{C-flooding} P_{coll} + P_{LL} P_{C-DSC} P_{coll} \tag{16}$$

Additionally, the rate of fatalities with respect to a total number of people on board is derived, which results in a number of fatalities per year. This, together with the probability for the loss of life, determined through Eq. (16), leads to a cumulative density function of the number of fatalities. The latter is called *F–N* and it is considered as an outcome of this framework.

In the following sub-sections, simplified models addressing accident response are introduced. Also an approach for modelling the rate of fatalities is shown.

4.13.1. The rate of fatalities and number of fatalities

In a case where the accident response is not effective (LL=1), the number of fatalities (N) is estimated. This parameter is one of the most relevant for the risk framework, and should be modelled as accurately as possible. However, as a result of a lack of information regarding the relationship between the assumed explanatory variables, namely *time to capsize*, *evacuation time*, *number of passengers on board the ship* and the response variable N, it is difficult to define a precise model predicting N. Therefore, conservative assumptions are adopted in this study. Such a choice can be justified by the aim of this paper, which is to introduce a framework for risk assessment, showing its abilities for efficient reasoning and instantaneous updating in the light of new knowledge, rather than delivering the "true numbers" for risk. Thus, N is assumed to be inversely proportional to the ratio *haz/resp* and the number of passengers on board as follows:

$$N = (1 - (haz/resp))N_{passengers} \tag{17}$$

where $1-(haz/resp)$ is considered as the rate of fatalities. Although this assumptions appears straightforward, the results obtained are comparable with the accident statistics, as presented in Fig. 9 and discussed in Section 5.4.

4.13.2. The time needed for evacuation of a RoPax

Assuming an ordered evacuation of a ship in danger, the time to evacuate the ship (TTE) is modelled with the use of triangular distributions following the IMO recommendations, and a distinction is made between day and night; see [73]. Additionally, an assumption is made regarding the effect of challenging weather conditions on the evacuation time, in a fashion presented in the following equation:

$$TTE = \begin{cases} T1, & Day\ and\ wave\ height < 3\text{ m} \\ T2, & Night\ and\ wave\ height < 3\text{ m} \\ T3, & Day\ and\ wave\ height > 3\text{ m} \\ T4, & Night\ and\ wave\ height > 3\text{ m} \end{cases} \tag{18}$$

where $T1 = [20, 20, 40]$, $T2 = [20, 40, 40]$, $T3 = [20, 40, 60]$, $T4 = [25, 40, 60]$ in minutes.

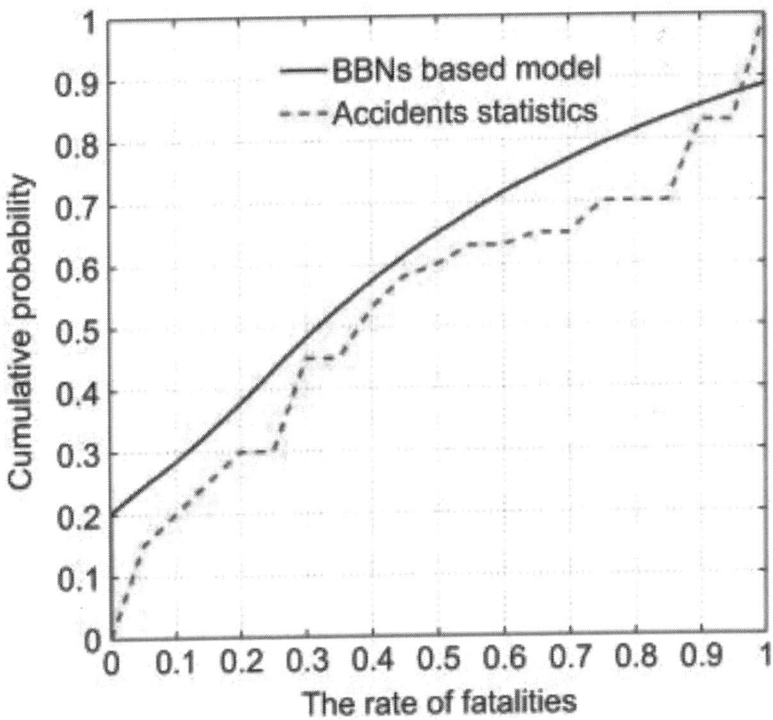

Figure 9. A CDF describing the rate of fatalities obtained from the model (solid line) plotted against the CDF reflecting historical data from 1987 to 2007 (dashed line).

4.13.3. The time needed for tugs

The time needed for tugs to arrive at the scene is based on the location of the accident relative to the nearest shore rescue station and possible weather conditions. The probable locations of an open sea collision in the GoF in which a RoPax is expected to be involved are obtained from the DMTS, and are depicted in Fig. 2. Thus, the time needed for tugs to reach the scene varies from 1 to 3 h, given good weather conditions. In the event of bad weather this time is increased by a factor of 1.5.

4.13.4. The number of passengers on board the ship

In the case study presented here, the number of passengers on board the ship during a collision (*Npassengers*) is modelled by means of triangular distribution, with a lower limit of 200, an upper limit of 3000 and mode value of 900. The lower limit is an estimate for a ship's crew and a minimal amount of passengers, whereas the upper limit is an estimate of the maximum capacity of RoPax ships cruising in the GoF. The mode value is based on the available data published by ship operators regarding the total monthly volume of passengers transported, see for example [74].

5. RISK FRAMEWORK VALIDATION

Once the framework structure and its parameters are defined and the results are obtained, we validate the framework by performing four analyses, to determine the following:

1 the sensitivity of the framework to the stochastic parameters (the sensitivity analysis);
2 the distribution of uncertainty across the framework (the value-of-information analysis);
3 the sensitivity of the framework to changes in the assumptions governing the framework (the influence analysis);
4 the agreement of the framework with the real-world conditions.

5.1. The sensitivity analysis

The sensitivity analysis is performed to identify the essential variables which have the highest impact on the outcome of the model. For this purpose, every conditional and prior probability in the BBNs is systematically varied in turn while keeping the others unchanged. This allows the effects on the output probabilities computed from the network to be examined, see for example [40].

To determine the sensitivity of an output variable to a given parameter of the model a sensitivity function is estimated for each single variable. This sensitivity function describes outcome of the model as a function of the parameter *z*, which takes the following form:

$$z = p(Y = y_i \mid \pi) \quad (19)$$

where *yi* is one state of a network variable *Y*, and π is a combination of states for *Y*s parent nodes. The sensitivity function takes a general form as follows, see [75]:

$$f(z) = (c_1 z + c_2)/(c_3 z + c_4) \quad (20)$$

where *f*(*z*) is the output probability of interest given observations, and $c_{1 \ldots 4}$ are the constants, which are identified based on the model. The first derivative of this function describes the effect of minor changes in a variable on the output and is called the sensitivity value. The sensitivity values were obtained using dedicated tools, which are implemented in the software package GeNIe, used for the development of BBNs presented here.

The results of the analysis indicate that the model is sensitive to changes in the following nodes: *stability conditions*, *collision mass ratio*, *collision probability*, *capsizing as a result of flooding* and *time to evacuate a ship*; see Table 6. In the case study presented here, we attempt to tailor these variables to the specific ship type and location, and evaluate them as accurately as possible using the methods described in the previous sections. However, certain variables could not be determined without ambiguity, e.g. *evacuation time*, and thus the effect of those on the output of the framework is determined in Section 5.3. The distributions for the nodes with a lesser impact on the outcome of the framework are based on the generic data available in the literature, natural laws and the authors' best judgement.

Table 6. Results of the sensitivity analysis.

Variable name	Maximum absolute sensitivity value
Stability conditions	0.02
Collision mass ratio	0.02
Collision probability	0.01
Capsizing as a result of flooding	0.002
Time to evacuate a ship	0.0001

5.2. The value-of-information analysis

The value-of-information analysis identifies the most informative variables, with respect to the output variable, determining the variables among which the probability mass of the output is scattered. For this purpose the concept of entropy is utilised, which is a measure of the randomness of a variable; the higher the randomness, the higher the entropy. The entropy $H(X)$ of a discrete random variable X, consisting of a number of states x, is defined as follows; see [76]:

$$H(X) = -\sum_{x} p(x) \lg p(x) \tag{21}$$

In a model, where the outcome variable is conditionally dependent on a number of parental variables, the conditional entropy $H(X|Y)$ needs to be applied. This is a measure of the uncertainty of X given an observation of Y and is estimated using the following formulae; see [76]:

$$H(X \mid Y) = \sum_{y} p(y) H(X \mid Y = y) = H(X) - I(X;Y)$$

$$I(X;Y) = \sum_{(x,y)} p(x,y) \lg \frac{p(x,y)}{p(x)p(y)} \qquad (22)$$

where $I(X;Y)$ is the mutual information, which explains the reduction in the entropy of a variable X by observing a variable Y. Thereby, the variable Y to be observed next is the one that has the maximum value of $I(X;Y)$. The results of the value-of-information analysis with respect to the model output are gathered in Table 7. They reveal that the uncertainties of the output variables are mostly explained by the distributions of the variables named *capsizing as a result of flooding*, *collision probability*, *stability conditions*, *time to evacuate a ship*, *inner hull rupture* and *collision mass ratio*. The remaining variables are less informative.

Table 7. Results of the value of information analysis.

Variable name	Mutual information
Capsizing as a result of flooding	0.60
Collision probability	0.30
Stability conditions	0.30
Time to evacuate a ship	0.15
Inner hull rupture	0.10
Collision mass ratio	0.06

However, neither of these two analyses can be performed on a model that contains either PPDs or a continuous variable. Thus, the variables needed to be discretized prior to the analyses.

5.3. The influence analysis

The two analyses above provide an insight into the given BBNs, comprising the given variables linked with certain likelihood functions (*LFs*). However, if these functions are determined ambiguously, as a result of unavailability of knowledge or limited understanding of a given phenomenon (e.g. *collision angle*) or limited resources (e.g.*damage extent significant, ships stay separated, evacuation time*), changes in the *LFs*may also affect the results of the sensitivity and value-of-information analysis. Thus, the influence analysis is performed, which aims at the quantification of the effect of changes in the assumptions on the outcome of the framework and other relevant variables. The proposed risk framework allows very effective influence analysis to be performed, as the updating process of BBNs is instantaneous. In the case study presented here, the following variables are subject to the influence analysis: *stability conditions, collision angle, collision angle significant, damage extent significant, ships stay separated,machinery damaged* and *time to evacuate a ship.*

The parameters and the various definitions of them adopted for the influence analysis are presented in Table 8. Then the framework is evaluated for the number of possible combinations of the given input parameters, and thus a set of results (F−N) is obtained. Then the medians and standard deviations of *F* for each *N* are determined, and a 95% confidence band around the mean value is obtained. The results of the analysis are depicted in Fig. 10.

Table 8. Influence analysis set-up.

Variable name		Definitions adopted
Stability conditions	s	s1; s2; s3; s4
Collision angle	α	Uniform(10,170); triangular(10,90,170)
Collision angle significant	α_{sign}	Uniform(45,135); uniform(20,160)
Damage extent significant	des	$1.2L_{struct}$; $1.4L_{struct}$
Ships stay separated	sss	$f(des_i, \alpha_{sign})$, where $i = 1,2,3$
Machinery damaged	md	0.1; 0.16; 0.25
Time to evacuate a ship	TTE	T1; T2; T3; T4 – see Section 4.13.2

5.4. The agreement of the framework with the real-world conditions

Once the framework is defined and the results are obtained, they are compared with the available data on the risk and severity of open-sea collisions in which RoPax ships were involved.

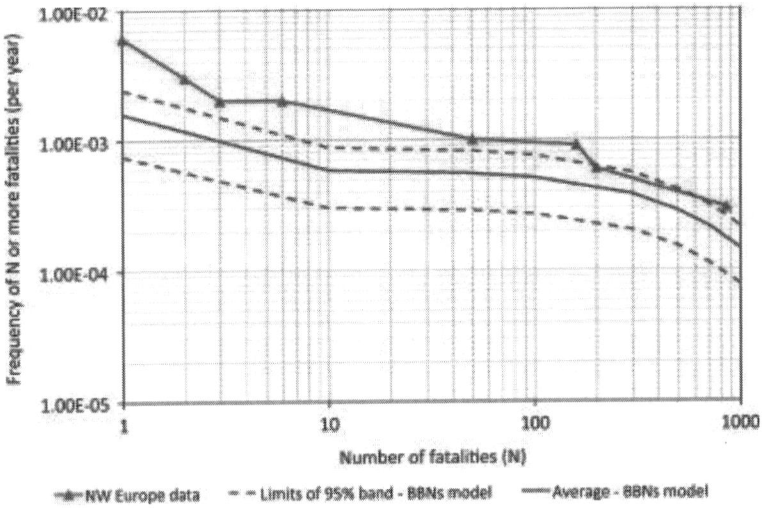

Figure 10. A risk picture obtained with the use of the risk framework developed in this paper, for the Gulf of Finland (in blue), plotted against the historical data (1978–1994) on the risk to RoPax ships in European waters (in red). (For interpretation of the references to color in this figure caption, the reader is referred to the web version of this paper.)

First, the *F–N* graph that was obtained is compared with the graph based on the accident statistics; see Fig. 10. The latter is obtained from the available accident data on RoPax ships for the North–West European seas and is adapted from [18]. The obtained *F–N*graph contains the accumulated values of the risk of a RoPax capsizing as a result of flooding or DSC, whereas the accident-based graph covers all kinds of accidents (collision, grounding, fire). In general, an *F–N* can be considered as a suitable risk picture reflecting the definition of risk as "an R set" given by Eq. (1), as it comprises the probability of various levels of consequences (*N*) along with the associated uncertainties expressed as a 95% confidence band. The results obtained show a good level of agreement of the modelled data with the observed data for higher numbers of fatalities. This can be explained by the fact that in the history of RoPax accidents, the most severe cases were those accidents associated with the ship capsizing as a result of the car deck flooding. However, none of the recorded capsizing accidents were caused by a collision with another ship. Moreover, it should be noted that the statistics-based F−N represents the risk of all types of accidents, not only collisions, which may be a reason for the divergence of plots for a smaller *N*.

Second, the following three intermediate quantities are validated with the available data presented in [7], [5], [13] and [77]:

1 the conditional probability of the loss of a RoPax as a result of flooding: $P_{C\text{-}flooding}$;
2 the probability of serious damage given an open sea collision: P_{des};
3 the rate of fatalities: $1-(haz/resp)$.

The probabilities that are obtained with the use of the risk framework presented here are continuous variables, following certain, non-parametric distributions. However, for the purpose of the validation and visualisation of the results, the average values are used where necessary; see Table 9. The rate of fatalities is taken as a distribution and compared with the distribution obtained from the accidents statistics.

Table 9. Risk framework verification, comparison of various models.

Parameter	Value	Limitations	Reference
$P_{C-flooding}$	1.80×10^{-2}	valid for the specific RoPax vessel and location of this study	
	1.76×10^{-2}	accidents statistics-based model	[5]
	1.89×10^{-2}	valid for the specific RoPax vessel, supported by accidents data	[7]
	4.20×10^{-2}	valid for the specific RoPax vessel and location (the Atlantic Ocean)	[13]
P_{des}	1.50×10^{-1}	valid for the specific RoPax vessel and location of this study	
	1.60×10^{-1}	accidents statistics-based model	[5]

The average value of $P_{C\text{-}flooding}$ obtained from the framework is close to the results obtained from the model based on global statistics, and to the results of analysis for the specific RoPax as presented in [5] and [7]. However, the value of $P_{C\text{-}flooding}$ delivered by the risk framework introduced here is different from the results for a specific RoPax operating in the Atlantic Ocean, as shown in [13], but still of the same order of magnitude. The average value of P_{des} is in line with the corresponding probability of serious damage given open sea collision, according to historical data, as compiled in [5].

Also, good agreement is found between two distributions describing the rate of fatalities, as depicted in Fig. 9. The distribution obtained from the data is discreet, as it is based on 31 cases, as compiled in [77]. The distribution obtained from the model is continuous. The agreement between these two can be seen especially for the rates between 0.3 and 0.5 and for the rate 0.9. According to the accidents statistics [77] the probability of the rate of fatalities falling in the range (0.95–1), given accidental flooding, is 0.15. The

BBNs-based model delivers a number of 0.16 for the same range. Otherwise, the model presented here tends to slightly overestimate this parameter for rates lower than 0.95. However, for precise calculation of this parameter, one may adopt models simulating ship evacuation processes, which need detailed description of ship interior, see for example [78],[79] and [80]. Another possibility is to use generic models, for instance [81] and [82], which require high-level description of the arrangement of escape routes onboard a ship. The results obtained from those models can be used for detailed validation of the simplified approach proposed here.

Despite the simplifications adopted in the framework, the results of the validation show that the presented BBNs-based model provides sound estimations for all three parameters. As these parameters are of high importance for the analyzed phenomena, the above findings provide a basis for beliefs about reliable output of the model. Ultimately the framework delivers risk predictors for the analyzed scenarios which are in good agreement with the available historical data.

6. CONCLUSIONS

This paper presents a framework for risk analysis and assessment in maritime transportation systems, following the formal requirements adopted and reflected by Eq.(1). The framework is systematic, proactive and transferable. It utilises BBNs as medium to express and propagate the background knowledge that is available about the system being analysed. The BBNs applied here combine discrete and continuous variables. This allows probabilistic relationships among the variables and complex dependencies as well as fast propagation of information through the framework. Moreover, the processes of incorporation of new knowledge or new data into the framework and the quantification of the effect of changes in the assumptions on the outcome are very efficient, as the BBNs allow for instantaneous updating. This, together with the sensitivity analysis and the value-of-information analysis, increases the credibility of the results which are obtained, as the distribution of knowledge, on which the framework is developed, is presented. This, in turn, may help in determining the areas of the framework which are either unknown or not understood well enough, and have significant effects on the outcome of the framework, meaning that they need to be treated with caution.

Then, the framework is used for a case study, evaluating the risk for an MTS operating in the GoF during the ice-free season considering an open-sea collision, where a RoPax is struck by another ship and the RoPax is lost as a result of hull rupture and consecutive flooding, or as a result of post-accidental propulsion failure and exposure to high waves, resulting in capsizing. This means that the other collision scenarios, such as a RoPax is a striking ship and suffers significant damages to her bow leading ultimately to water ingress, are not considered here. Moreover, the distribution of mass ratios of striking and struck ships accounts for 80% of traffic in the GoF. The cases where the mass of a striking ship is larger than 1.7 times that of a struck ship, which accounts for less than 5% of the traffic in the area, are not considered in the numerical analysis. However, these ratios are included in the risk framework, and their effects are conservatively estimated; extrapolating findings for other ratios.

Additionally, four stability conditions of a RoPax that is struck are anticipated, and it is assumed that they are equally likely. This needs further investigation, as this is a sensitive parameter of the framework introduced.

The results are valid within certain, predefined boundary conditions, but the modular nature of the risk framework allows its continuous improvement and adaptation to various conditions. Despite the simplifications and the assumptions made in the framework, the results obtained are promising, and they show good agreement with the available statistical or modelled data on RoPax accidents operating around Europe.

However, this paper does not claim any "true numbers" for risk in the Gulf of Finland, as the aim of the analysis presented is to demonstrate the abilities of the framework for knowledge-based risk assessment and proper reflection of uncertainties which are inherent to any risk analysis. Further studies are desirable in order to cover missing accident scenarios (e.g. grounding or fires on board RoPax) and to improve the areas of the framework which are relevant, but about which proper background knowledge is lacking (e.g. the number of fatalities resulting from a collision, the extent of the damage resulting in catastrophic flooding or the collision speed and angle). Additionally, the framework allows the effect of risk control options to be studied. Both direction reasoning make it possible to propagate the outcome backwards, which can be utilised in a recommendation phase, pointing out the most effective ways to make improvements. This is, however, beyond the scope of this paper.

Nevertheless, the risk framework introduced here can be considered as a contribution to a concept of holistic risk modelling which can be utilised at the stages of either ship design or operation.

ACKNOWLEDGMENTS

The authors appreciate the financial contributions of the following entities: the EU, as this research was co-funded by the FAROS project (2012–2015), the city of Kotka, the Finnish Ministry of Employment and the Economy, and the Norwegian Ministry of Trade and Industry.

Furthermore, the authors are also indebted to the anonymous reviewers and the editor whose comments improved the presentation and the content of an earlier version of the manuscript. Hannah Ploskonka is appreciated for checking this paper for English.

The probabilistic models introduced in this paper were created using the GeNie modelling environment developed at the Decision Systems Laboratory, University of Pittsburgh available from http://genie.sis.pitt.edu/.

REFERENCES

1. Mains C. Updated damage statistics on collision and grounding. Report 1-11- D-2001-01-1, Germanisher Lloyd, 2001.
2. Papanikolaou A, Eliopoulou E. On the development of the new harmonised damage stability regulations for dry cargo and passenger ships. Reliab Eng Syst Saf 2008;93:1305–16.
3. Konovessis D, Vassalos D. Risk evaluation for RoPax vessels. Proc Inst Mech Eng Part M: J Eng Maritime Environ 2008;222:13–26.
4. Mermiris G, Konovessis D, Vassalos D. First-principles collision risk analysis of a RoPax vessel. In: Proceedings of 4th international ASRANet colloquium, 2008.
5. Guarin L, Konovessis D, Vassalos D. Safety level of damaged RoPax ships: risk modelling and cost-effectiveness analysis. Ocean Eng 2009;36:941–51.

6. Papanikolaou A, Mains C, Rusås S, Szalek R, Tsakalakis N, Vassalos D, et al. Goalds - goal based damage stability. In: Proceedings of the 11th international ship stability workshop. Wageningen: MARIN; 2010. p. 46–57.

7. Konovesis D, Vassalos D. Risk-based design for damage survivability of passenger ro-ro vessels. Int Shipbuild Prog 2007;54:129–44.

8. Papanikolaou AE. Risk-based ship design methods, tools and applications. Berlin, Heidelberg: Springer; 2009.

9. Santos TA, Guedes Soares C. Numerical assessment of factors affecting the survivability of damaged ro-ro ships in waves. Ocean Eng 2009;36:797–809.

10. Spanos D, Papanikolaou A. On the time dependent survivability of RoPax ships. In: Proceedings of the 11th international ship stability workshop. Wagenin- gen:MARIN; 2010. p. 143–7.

11. IMO. Guidelines for formal safety assessment (FSA) for use in the IMO rule- making process. MSC/Circ.1023; MEPC/Circ.392, 2002.

12. Vanem E, Puisa R, Skjong R. Standardized risk models for formal safety assessment of maritime transportation. In: ASME 28th international confer- ence on ocean, offshore and arctic engineering (OMAE 2009). 2009. p. 51–61.

13. Otto S, Pedersen PT, Samuelides M, Sames PC. Elements of risk analysis for collision and grounding of a roro passenger ferry. Mar Struct 2002;15:461–74.

14. Vanem E, Skjong R. Collision and grounding of passenger ships - risk assessment and emergency evacuations. In: Snaj, editor. Proceedings of the 3rd international conference on collision and grounding of ships, 2004. p. 195–202.

15. Kobylinski L. Stability and safety of ships: holistic and risk approach. Reliab Risk Anal: Theory Appl 2008;1:95–105.

16. Gerigk M. A method of risk and safety assessment during the ship salvage using the hazard, release and consequence analysis. J KONBiN 2010;13: 165–176.

17. Antao P, Guedes Soares C. Fault-tree models of accident scenarios of RoPax vessels. Int J Autom Comput 2006;3:107–16.

18. Konovessis D, Vassalos D, Mermiris G. Risk analysis for RoPax vessels. WMU J Maritime Affairs 2008;7:109–31.

19. Martins MR, Maturana MC. Application of Bayesian belief networks to the human reliability analysis of an oil tanker operation focusing on collision accidents. Reliability Engineering & System Safety 2013;110:89–109.

20. Groth K, Wang C, Mosleh A. Hybrid causal methodology and software platform for probabilistic risk assessment and safety monitoring of socio-technical systems. Reliab Eng Syst Saf 2010;95:1276–85.

21. Trucco P, Cagno E, Ruggeri F, Grande O. A Bayesian belief network modelling of organisational factors in risk analysis: a case study in maritime transportation. Reliab Eng Syst Saf 2008;93:845–56.

22. Vinnem J, Bye R, Gran B, Kongsvik T, Nyheim O, Okstad E, et al. Risk modelling of maintenance work on major process equipment on offshore petroleum installations. J Loss Prev Process Ind 2012;25:274–92.

23. Gran B, Bye R, Nyheim O, Okstad E, Seljelid J, Sklet S, et al. Evaluation of the risk OMT model for maintenance work on major offshore process equipment. J Loss Prev Process Ind 2012;25:582–93.

24. Leva, M. C., Friis-Hansen, P., Ravn, E. S., & Lepsoe, A. (2006). SAFEDOR: A practical approach to model the action of an officer of the watch in collision scenarios. In European safety and reliability conference; Safety and reliability for managing risk; ESREL 2006 (pp. 2795–2804). Francis, London.

25. Grabowski M, Merrick JRW, Harrold JR, Massuchi TA, van Dorp JD. Risk modeling in distributed, large-scale systems. IEEE Trans Syst Man Cybern Part A: Syst Hum 2000;30:651–60.

26. Szwed P. Risk factors and theory building: a study to improve passenger vessel safety. WMU J Maritime Affairs 2011;10:183–208.

27. Madsen A, Lang M, Kjrulff U, Jensen F. The hugin tool for learning Bayesian networks. In: Nielsen T, Zhang N, editors. ECSQARU 2003, LNAI 2711. Berlin, Heidelberg: Springer-Verlag; 2003. p. 594–605.

28. Darwiche A. Modeling and reasoning with Bayesian networks. 1st ed.Cambridge University Press, New York, USA; 2009.

29. Drużdzel M, van der Gaag L. Building probabilistic networks: "Where do the numbers come from?" guest editors' instruction IEEE Trans Knowl Data Eng 2000;12:481–6.

30. Montewka J, Weckström M, Kujala P. A probabilistic model estimating oil spill clean-up costs – a case study for the gulf of finland. Mar Pollut Bull 2013;76:61–71.

31. Zio E, Apostolakis G. Two methods for the structured assessment of model uncertainty by experts in performance assessments of radioactive waste repositories. Reliab Eng Syst Saf 1996;54:225–41.

32. Kaplan S. The words of risk analysis. Risk Anal 1997;17:407–17.

33. Ulusçu ÖS, Özbaş B, Altıok T, Or İ. Risk analysis of the vessel traffic in the strait of Istanbul. Risk Anal 2009;29:1454–72.

34. van Dorp JR, Merrick JRW. On a risk management analysis of oil spill risk using maritime transportation system simulation. Ann Oper Res 2011;187:249–77.

35. Stirling A, Gee D. Science, precaution, and practice. Public Health Rep 2002;117:521–33.

36. Montewka J, Goerland F, Kujala P. On a risk perspective for maritime domain. J Pol Saf Reliab Assoc 2013;4:101–8.

37. Myllymäki P, Silander T, Tirri H, Uronen P. B-course: a web-based tool for Bayesian and causal data analysis. Int J Artif Intell Tools 2002 369–87.

38. Kontkanen, P., Myllymaki, P., Silander, T., & Tirri, H. (1997). Comparing predictive inference methods for discrete domains. In Sixth International Workshop on Artificial Intelligence and Statistics (pp. 311–318).

39. Uusitalo L. Advantages and challenges of Bayesian networks in environmental modelling. Ecol Model 2007;203:312–8.

40. Hänninen M, Kujala P. Influences of variables on ship collision probability in a Bayesian belief network model. Reliab Eng Syst Saf 2012;102:27–40.

41. Merrick JR, Van Dorp JR. Speaking the truth in maritime risk assessment. Risk Anal 2006;26:223–37.

42. Goerlandt F, Ståhlberg K, Kujala P. Influence of impact scenario models on collision risk analysis. Ocean Eng 2012;47:74–87.

43. Aven T. A conceptual framework for linking risk and the elements of the data- information-knowledge-wisdom (DIKW) hierarchy. Reliab Eng Syst Saf 2013;111:30–6.

44. Pedersen PT. Review and application of ship collision and grounding analysis procedures. Mar Struct 2010;23:241–62.

45. Goerlandt F, Kujala P. Traffic simulation based ship collision probability modeling. Reliab Eng Syst Saf 2011;96:91–107.

46. Li S, Meng Q, Qu X. An overview of maritime waterway quantitative risk assessment models. Risk Anal 2012;32:496–512.

47. Mazaheri A, Montewka J, Kujala P. Modeling the risk of ship grounding—a literature review from a risk management perspective. WMU J Maritime Affairs http://dx.doi.org/10.1007/s13437-013-0056-3, 2013;12.

48. Montewka J, Krata P, Goerlandt F, Kujala P. Marine traffic risk modelling—an innovative approach and a case study. Proc Inst Mech Eng Part O: J Risk Reliab 2011;225:307–22.

49. Montewka J, Ehlers S, Tabri K. Elements of risk analysis for LNG tanker maneuvering with tug assistance in a harbor. In: 11th international sympo- sium on practical design of ships and other floating structures, vol. 2. Rio de Janeiro, Brazil: PRADS. 2010, p. 1563–72.

50. Apostolakis G. The concept of probability in safety assessments of technolo- gical systems. Science 1990;250:1359–64.

51. Aven T, Reniers G. How to define and interpret a probability in a risk and safety setting. Saf Sci 2013;51:223–31.

52. Winkler RL. Uncertainty in probabilistic risk assessment. Reliab Eng Syst Saf 1996;54:127–32.

53. Langseth H, Portinale L. Bayesian networks in reliability. Reliab Eng Syst Saf 2007;92:92–108.

54. Vanem E, Rusås S, Skjong R, Olufsen O. Collision damage stability of passenger ships: holistic and risk-based approach. Int Shipbuild Prog. 2007;54:323–37.

55. Aven T, Heide B. Reliability and validity of risk analysis. Reliab Eng Syst Saf 2009:1862–8.

56. Paté-Cornell ME. Uncertainties in risk analysis: six levels of treatment. Reliab Eng Syst Saf 1996;54:95–111.

57. Laboratory DS. GeNIe Documentation. University of Pittsburgh. Availalble from: ⟨http://genie.sis.pitt.edu/wiki/GeNIe_Documentation⟩, 2011.

58. Drużdzel M. Genie: A development environment for graphical decision- analytic models. In: Association AMI, editor. Proceedings of the annual symposium of the American Medical Informatics Association (AMIA), 1999, Washington, DC. p. 1206.

59. Kujala P, Hanninen M, Arola T, Ylitalo J. Analysis of the marine traffic safety in the gulf of finland. Reliab Eng Syst Saf 2009;94:1349–57.

60. Lützen M. Ship collision damage [Ph.D. thesis]. Technical University of Den- mark, Department of Mechanical Engineering, 2001.

61. Weng J, Meng Q, Qu X. Vessel collision frequency estimation in the singapore strait. J Navig 2012;65:207–21.

62. Tabri K. Dynamics of ship collisions [Ph.D. thesis]. Aalto University, School of Science and Technology, Espoo, Finland, 2010.

63. Ehlers S, Tabri K. A combined numerical and semi-analytical collision damage assessment procedure. Mar Struct 2012;28:101–19.

64. Ehlers S. Material relation to assess the crashworthiness of ship structures [Ph. D. thesis]. Aalto University, School of Science and Technology, Espoo, Finland, 2010.

65. Ehlers S. The influence of the material relation on the accuracy of collision simulations. Mar Struct 2010;23:462–74.

66. Ståhlberg K, Goerlandt F, Ehlers S, Kujala P. Impact scenario models for probabilistic risk-based design for ship–ship collision. Mar Struct 2013;33:238–64.

67. Papanikolaou A. Benchmark study on the capsizing of a damaged ro-ro passenger ship in waves. Issued for comments to: the ITTC specialist committee on the prediction of extreme motions and capsizing, 2001.

68. Papanikolaou A, Spanos D. Benchmark study on numerical codes for the prediction of damage ship stability in waves. In: Proceedings of the 10th international ship stability workshop. Hamburg: Germanischer Lloyd; 2007.

69. Schreuder M, Hogström P, Ringsberg J, Johnson E, C.-E. J. A method for assessment of the survival time of a ship damaged by collision. J Ship Res 2011;55:86–99.

70. BMT Ltd, Global Wave Statistics, BMT Ltd, London, 1986.

71. BureauVeritas, Rules for the classification of steel ships. Technical Report, Bureau Veritas; 2005.

72. Matusiak J. On the non-linearities of ships restoring and the Froude–Krylov wave load part. Int J Naval Archit Ocean Eng 2011;3:111–5.

73. IMO, Guidelines for evacuation analysis for new and existing passenger ships, MSC.1/Circ.1238, 2007.

74. TALLINK, Tallink exceeded the milestone of 900 000 passengers in June, 2013. [75] Chan H, Darwiche A. When do numbers really matter? J Artif Intell Res 2002;17:265–87.

75. Hugin Experts A/S, HUGIN API reference manual. Technical Report Version 7.4, Hugin; 2010. Available online: ⟨http://download.hugin.com/webdocs/man uals/api-manual.pdf⟩.

76. Jasionowski A, Vassalos D, Scott A. Ship vulnerability to flooding. In: 3rd international maritime conference on design for safety, Berkeley, California, 2007.

77. Ginnis AI, Kostas KV, Politis C, Kaklis PD. Velos: a VR platform for ship-evacuation analysis. Comput Aided Des 2010;42:1045–58.

78. Roh M-I, Ha S. Advanced ship evacuation analysis using a cell-based simula- tion model. Computers in Industry 2013;64(1):80–9.

79. Vanem E, Skjong R. Designing for safety in passenger ships utilizing advanced evacuation analyses—a risk based approach. Saf Sci 2006;44:111–35.

80. Łozowicka D. Organization of evacuation from passenger ships–a concept of safety enhancement. Zesz Naukowe 2012;32:110–4.

81. Łozowicka DH. Problems associated with evacuation from the ship in case of an emergency situation. Int J Autom Comput 2006;3:165–8.

CHAPTER 5

Impact Response of Aluminum Foam Sandwiches for Light-Weight Ship Structures

Vincenzo Crupi *, Gabriella Epasto and Eugenio Guglielmino

Department of Industrial Chemistry and Materials Engineering, University of Messina, C/da di Dio, Messina 98166, Italy

ABSTRACT

The structures realized using sandwich technologies combine low weight with high energy absorbing capacity, so they are suitable for applications in the transport industry (automotive, aerospace, shipbuilding industry) where the "lightweight design" philosophy and the safety of vehicles are very important aspects. While sandwich structures with polymeric foams have been applied for many years, currently there is a considerable and growing interest in the use of sandwiches with aluminum foam core. The aim of this paper was the analysis of low-velocity impact response of *AFS* (aluminum foam sandwiches) panels and the investigation of their collapse modes. Low velocity impact tests were carried out by a drop test machine and a theoretical approach, based on the energy balance model, has been applied to investigate their impact behavior. The failure mode and the internal damage of the impacted*AFS* have also been investigated by a Computed Tomography (*CT*) system.

KEYWORDS

aluminum foam sandwich; low velocity impact; Computed Tomography; shipbuilding

1. INTRODUCTION

The "lightweight design" philosophy is essential in the transport industry (automotive, aerospace, shipbuilding industry); lightweight structures are commonly used to increase payload, to reach higher speed and to obtain a lower fuel consumption. Thus, new materials and technologies have been adopted and, in particular, the use of sandwich structures has steadily increased in the last years. In order to meet the ever growing request of lighter vehicles and ships, the transport industry is engaged in the design and construction of structures quite different from the traditional ones, and as a consequence, a lack of experiences concerning the strength of many structural details has become apparent. A wide use of sandwich structures requires that all differences respect to the traditional ones be adequately accounted for. Thus, there is a need for new research activities in order to provide the designers of sandwich structures with new reliable data.

Moreover, problems of collision and crashing are very important for high speed terrestrial and marine vehicles and sandwich structures have so far shown good capabilities in absorbing energy, so it is necessary to acquire a better knowledge concerning the impact behavior of structural details built up resorting to sandwich technologies.

Core deformation and failure are decisive factors for the energy absorption capability of sandwich structures. After fracture of the skin, the impacting object may damage and penetrate into the core. With aluminum honeycomb cores, damage consists of crushing or "buckling" of cell walls in a region surrounding the impact point, while, in foam cores, damage looks more like a crack for low-energy impacts [1].

While polymeric foams have been applied for many years, currently there is a significant and growing interest in metal foams for transport applications. Aluminum foam sandwiches (*AFS*) [2,3], obtained by combining metal face sheets with a lightweight metal foam core, have peculiar properties (low specific weight, efficient capacity of energy dissipation, high impact strength, acoustic and thermal insulation, high damping), that made them interesting for a number of practical applications, such as the realization of lightweight structures with high mechanical strength and good capacity of energy dissipation under impacts. Aluminum sandwich structures are suitable for applications in high speed marine and terrestrial vehicles, as they allow a speed increase with a good passenger comfort thanks to their specific weight and high damping capacity. The use of lightweight materials is an important issue for ships and, in comparison with cars, a high flexibility of materials processing is needed, because ship are not realized in large series with highly standardized components, so aluminum foams or panels have a lot of advantages for possible applications in ship construction [4]. Closed cell metal foams are being developed for lightweight ship components such as elevator platforms, water-tight doors, hatches and bulkheads as declared by the Fraunhofer USA Center for Manufacturing and Advanced Materials, Delaware, USA.

Aluminum foams represent an attractive and interesting material for the development of highly efficient energy absorbers thanks to their capacity to undergo large strain at almost constant stress [5]. Foams can absorb a large amount of mechanical energy when they are deformed due to their high porosity, while stresses are limited to the compressive strength of the material, so they can act as impact energy absorbers which limit accelerations in crash situations. Foamed organic material has low specific weight, but the energy amount convertible to strain energy is relatively low for the low strength of the polymeric foams. As metal foams can have much higher collapse strengths than polymer-based foams, they can find applications in areas not accessible to foams up to date.

Low velocity impact tests were carried out on a range of polymeric foam sandwiches by Hazizan *et al.* [6] and an energy balance model was applied to predict their low velocity impact response.

In a previous research paper of the authors [7], the structural response of aluminum foam sandwiches under static loading was compared with that of the PVC foam sandwiches.

The failure mode and the damaged structure of the impacted panels have been also investigated by a Computed Tomography (*CT*) system, which allows a three-dimensional reconstruction of the analyzed object [8]. This non-invasive technique have been used to quantitatively characterize the microstructure and the internal architecture of different typologies of closed cell aluminum alloy foam [9] and to obtain the data for Finite Element models of open-cell aluminum foam specimens [10].

Aim of the present research was the analysis of low-velocity impact response of *AFS* panels and the investigation of their collapse modes. Two types of *AFS* panels were analyzed; they differ for the foam quality and for the skin-core adhesion. Low velocity impact tests were carried out at different impact velocities by a drop test machine in order to investigate and compare the structural response of the two *AFS* types in terms of energy absorption capacity. The collapse mode and the internal damage of the impacted panels have been investigated using 3D Computed Tomography. A theoretical approach, based on the energy balance model, has been applied to investigate their impact behavior and the model parameters were obtained directly from the measurements carried out on the tomographic images of the impacted sandwiches.

This combined experimental and theoretical study has particular importance for applications that require lightweight structures with a high capacity of energy dissipation, such as the transport industry.

2. MATERIALS AND METHODS

2.1. Materials

Two different commercial aluminum alloy foam sandwiches have been investigated (Figure 1): the first one (*Schunk-Honsel Entwicklungsgemeinschaft*) with faces obtained by extrusion (integral skins), the second one (*Alulight® International GmbH*) with faces bonded to the core by an epoxy adhesive.

Figure 1. Aluminum foam sandwiches (AFS) panels.

The chemical composition, physical and geometrical properties of the investigated AFS panels are reported in Tables 1and 2.

Table 1. Chemical composition of the *AFS* panels.

	AFS-Alulight	*AFS-Schunk*
Skin	Al 99.6%, O 0.4%	Al 98.97%, Mn 1.03%
Core	Al 82.3%, Si 17%, Ti 0.14%, O 0.56%	Al 92.83%, Si 7.03%, Ti 0.14%
Epoxy adhesive	C 72.12%, O 27.71%, Cl 0.17%	

Table 2. Physical and geometrical properties of the *AFS* panels.

	AFS-Alulight		*AFS-Schunk*	
	Skin	**Core**	**Skin**	**Core**
material	Al (99.5%)	AlSi10	AlMn1	AlSi7
density [kg/dm^3]	2.73	0.53 ± 0.06	2.73	0.45 ± 0.04
thickness [mm]	1	9	1	9
total density of *AFS* panel [kg/dm^3]	0.95 ± 0.05		0.87 ± 0.04	
total thickness of *AFS* panel [mm]	11		11	

2.2. Methods

The low-velocity impact tests were carried out by means of the *Ceast Fractovis Plus* drop test machine (Figure 2), able to eliminate multiple impacts. The mass of the impactor and the drop height are variable, allowing for a wide range of impact energies.

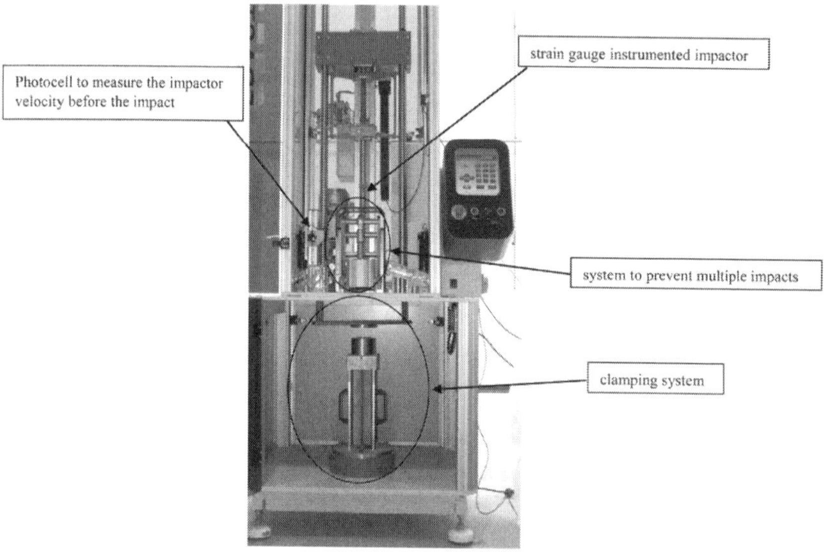

Figure 2. Drop-weight impact test machine.

The impacted damage of the aluminum foam core, undetectable by visual inspection, was analyzed by the *3D* Computed Tomography System *Y.CT Vario* (Figure 3).

This unit is equipped with an X-ray source having maximum voltage and current of 225 kV and 7.1 mA, respectively, depending on the focal spot size that can be chosen among these values: 250 μm, 300 μm, 500 μm and 800 μm. The detector system is a flat panel with a resolution of 1920 × 1536 pixels. The scans, reported in this paper, were conducted with 250 μm focus and X-rays were set at a voltage of 210 kV and at a current of about 1.1 mA. A conical X-ray beam scanned the sample, which was rotated at increments of 0.0087 rad/s for each rotation step. This procedure was then repeated until a full rotation of 2π rad was achieved, and a total of 1440 projections were then obtained to be used in the 3D profile generation. The voxels have cubic shape with edge length in the range between 0.033 to 0.050 mm and the image size is 2048 × 2048 pixels. The integration time was chosen equal to 500 ms. It is important to underline that this NDT technique does not require the cutting and polishing of the samples to carry out the X-ray measurements. This allows significant savings in time and the investigation of the internal damage without perturbing the impacted specimen. The system, based on a variable focal-spot size technology, creates the cross-sectional images of three-dimensional objects using X-rays. A volumetric representation of the item to be inspected is obtained as a result of the *CT*. Both the material inner and outer structures and the geometric dimensions of the item to be inspected are recognizable.

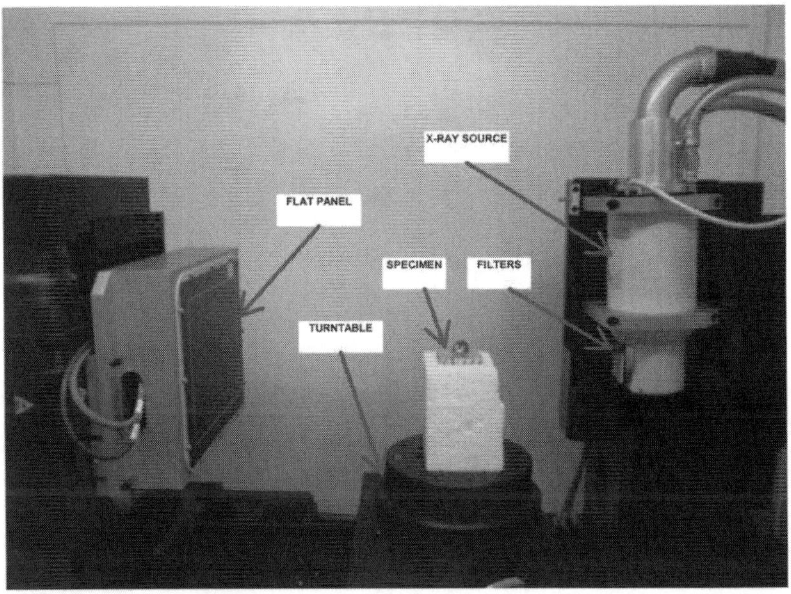

Figure 3. Computed Tomography system.

Figure 4 shows the *3D* reconstruction of an *AFS Schunk* panel by means of the *CT*. This non-destructive technique allows the investigation of the foam porosity and the localization and quantification of the pores; the analysis, in terms of grey levels intensity, along the middle line, drawn on the sample middle section, is reported in Figure 4.

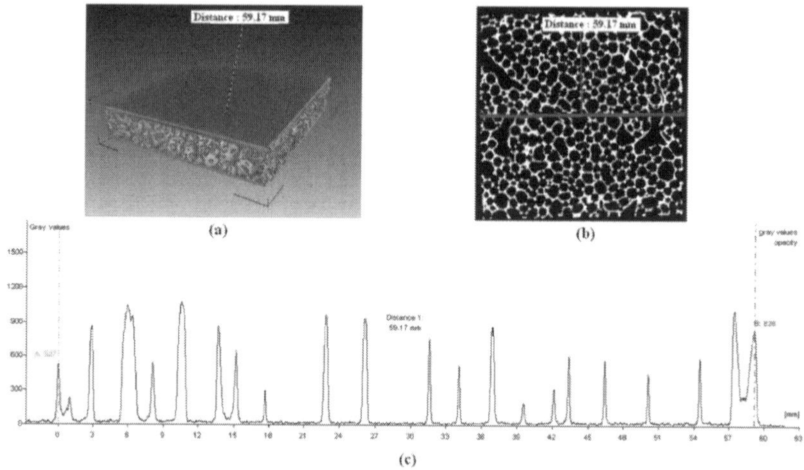

Figure 4. Computed Tomography (*CT*) analysis of an *AFS Schunk* panel.

Because of the non-homogeneity of the *AFS* panels, it is strongly necessary to check the sample before carrying out the impact test. It was decided to discard the panels which showed the following defect types in their whole volume: a cell crossing the whole section along the thickness, a cell whose largest dimension is greater than 5 times the average cell size or greater than 15 mm, as the one shown in Figure 5. The percentage of discarded samples was lower than 10%.

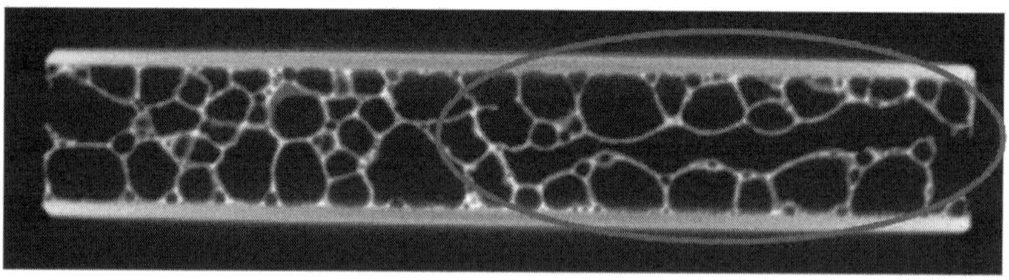

Figure 5. *CT* analysis of an *AFS Schunk* panel before the impact test.

3. EXPERIMENTAL SECTION

3.1. Low-Velocity Impact Tests

Dynamic impact tests were performed on specimens of the two *AFS* typologies (Schunk and Alulight) with an impactor mass of about 7 kg and different values of impact velocity ranging from 1.5 to 8 ms^{-1}. The impact energy values range from 8 to 225 J. The impactor, having a hemispherical tip with diameter of 20 mm, is instrumented by means of strain gauge, which allows the measurement of a force value until 40 kN (Figure 6). The *AFS* specimens (60 × 60 × 11 mm, foam core thickness 9 mm and skin thickness 1 mm) were fully fixed by a rigid metallic plate with a diameter of 40 mm, as shown inFigure 7, without crushing the sample.

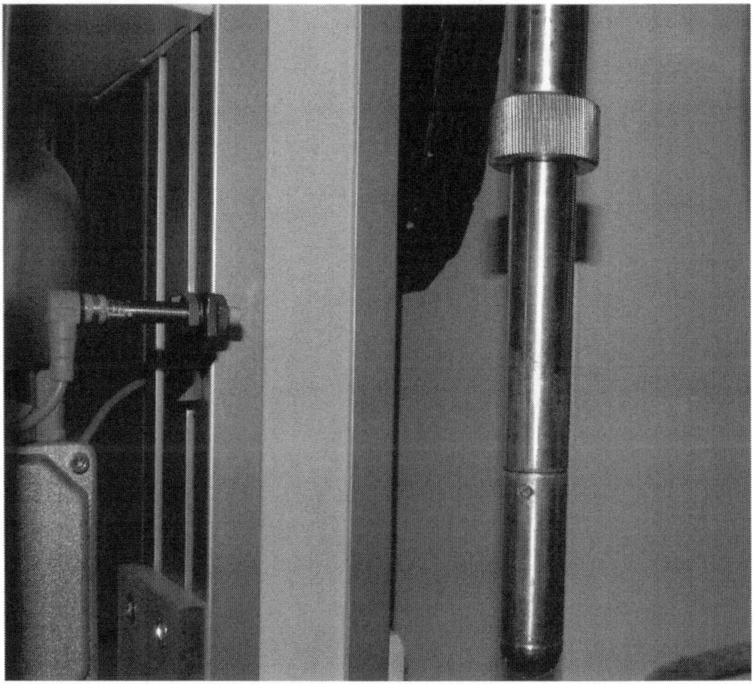

Figure 6. Impactor.

Figure 7. Clamping system.

The impact force was measured using the strain gauge mounted in the impactor of the drop test machine and the variation of this signal with time was recorded by a dedicated computer and analysed by the software. The velocity during impact is calculated by the integration of the acceleration over time, obtained by the impact force divided by the mass of the dropweight, then the displacement signal is obtained by means of the integration of the displacement.

The measured load-time curves at different impact velocities (v = 4, 5, 6, 7, 8 ms^{-1}), corresponding to *AFS* Schunk and Alulight specimens, are given in Figures 8 and 9.

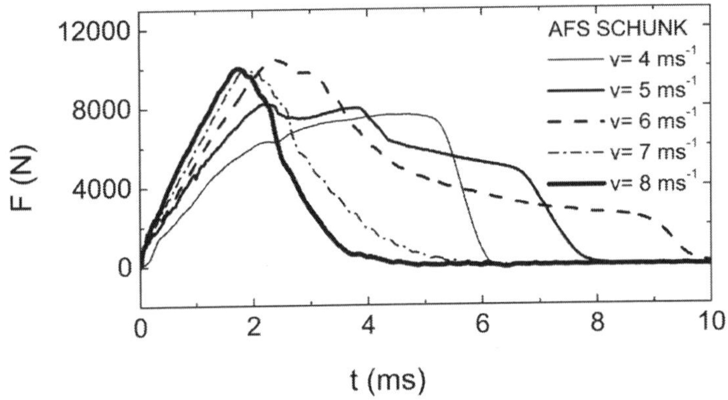

Figure 8. Load-time curves (*AFS Schunk*).

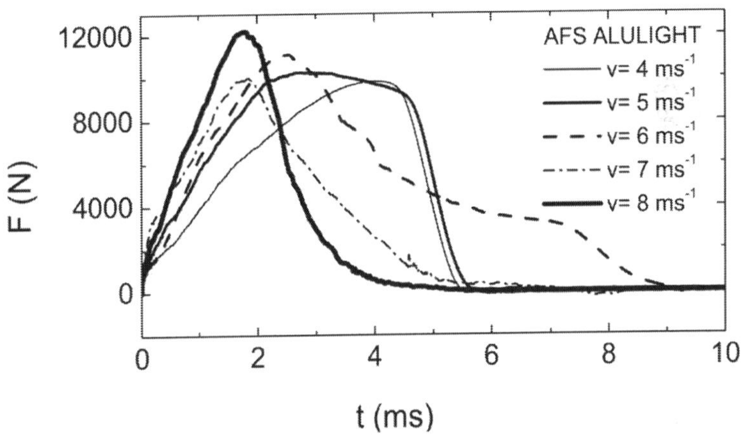

Figure 9. Load-time curves (*AFS Alulight*).

The contact force peak—velocity curves, obtained by all the tests performed on the two *AFS* typologies, are shown inFigure 10.

The data scattering, observed in the tests, is due to the different porosity distributions of the aluminum foams, so it is important to check the foam quality by means of non-destructive techniques, such as the *CT*. Table 3 reports the results of all the experimental tests in terms of absorbed energy and contact force peak. The energy amount, required to produce the complete failure of the sandwiches, was evaluated equal to 142 J for *AFS Alulight* panels and 122 J for *AFS Schunk* panels. The experimental results confirm the better response of *AFS Alulight* under impact loading even if they have a worse skin-core adhesion respect to the *AFS Schunk*, taking into account that there is no an high difference of core density between the two *AFS* types. Thus, the dynamic response of these sandwiches is not influenced by the skin-core adhesion and depends on the quality and mechanical properties of the foam core as confirmed by Mahfuz *et al.* [11], who demonstrated that a significant portion of the initial impact energy is absorbed by the core material.

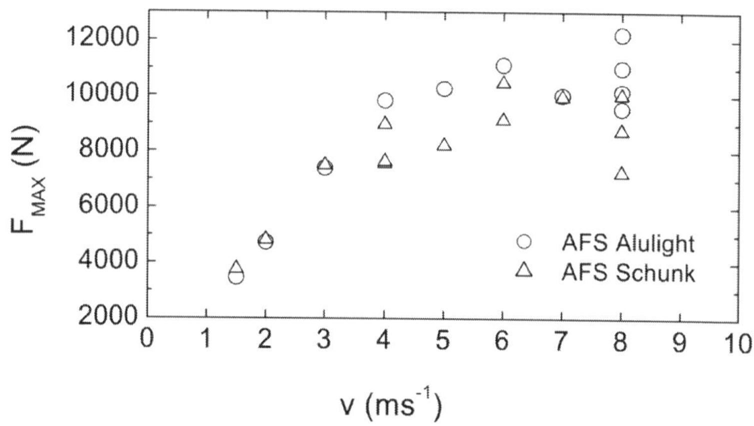

Figure 10. Contact force peak—velocity curves for *AFS* panels.

Table 3. Results of all the impact tests.

	AFS-Alulight		*AFS-Schunk*	
v [m/s]	F_{max} [N]	E [J]	F_{max} [N]	E [J]
1.5	3,458	8	3,757	8
2	4,722	14	4,810	14
3	7,398	31	7,499	31
4	9,822	56	8,967	56
4	-	-	7,567	56
4	-	-	7,640	56
5	10,257	88	8,214	88
6	11,112	127	10,460	127
6	-	-	9,132	127
7	10,010	139	9,948	130
8	11,010	144	8,769	119
8	9,542	134	8,740	117
8	10,150	137	10,010	134
8	12,213	158	7,253	110

Figure 11 shows the *AFS* panels after the impact at different impact velocities (v = 4, 5 and 6 m/s), confirming that the out-of-plane displacements are greater for *AFS Schunk* panels.

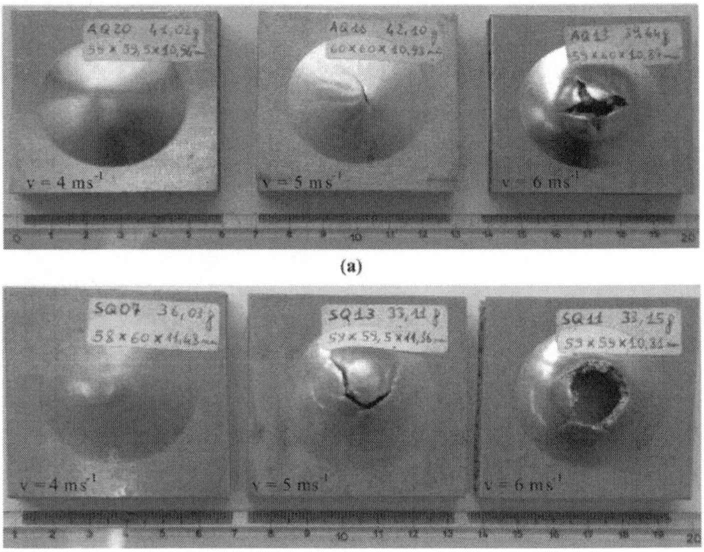

(b)

Figure 11. *AFS* panels ((**a**) *Alulight*; (**b**) *Schunk*) after the impact (v = 4, 5 and 6 ms^{-1}).

Figure 12 shows the tomographic images just for *AFS Alulight* and *Schunk* panels after the impact at v = 4 m/s. These tomographic images allow a better understanding of their failure mode, which is characterized by the progressive crushing of the foam cells with a more uniform distribution of the impacted load.

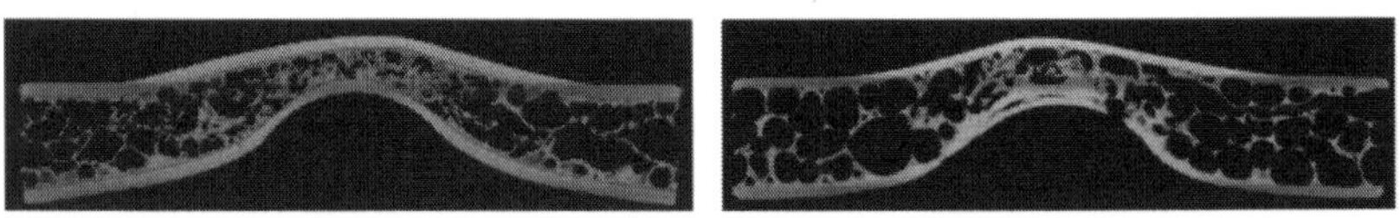

(a) (b)

Figure 12. CT images of the *AFS* panels ((**a**) *Alulight*; (**b**) *Schunk*) after impact (v = 4 m/s).

The post-impact investigation of the specimens confirms the results of the tests conducted by Compston *et al.* [12]; the *AFS*specimens experienced extensive ductile fracture with large out-of-plane displacement compared to the predominantly elastic behavior of the polymeric sandwiches; as a consequence the impact damage can be often detectable by a simple visual inspection of the sandwich structures. Moreover, the *AFS* structures are relatively intact compared to the more catastrophic and localized fracture of the polymeric sandwiches, so they exhibit a better post-impact damage tolerance and mechanical properties [7,8,12]. Thus, the *AFS* panels are a better choice with respect to the sandwich panels to realize progressive energy absorbers.

3.2. Energy Balance Model

The impact response of the sandwich structures was modeled using a theoretical approach, based on the energy-balance model [1] in order to examine the relative effects of the bending, shear and indentation components of the deformation. The sandwich beam was modeled as a combination of two springs (Figure 13), according to the model proposed by Shivakumar*et al.* [13]: a linear spring K_{bs} to account for the global deflection w_b and a nonlinear spring K_i to represent the local indentation effects.

The sandwich panel, clamped around its edges, experiences the local displacement α, due to the indentation of the top face and to the core crushing, and the global displacement w_b due to the bending and the shear of the entire panel.

Although the impact event is a highly dynamic event, statistically determined contact laws can be used in the impact dynamics analysis of low-velocity impacts because strain rate and wave propagation effects are negligible with commonly used material systems [1]. It was assumed that the target responds quasi-statically during the impact event and that all the initial kinetic energy has been absorbed. This assumption is realistic in the investigated cases because the drop test machine (Figure 2) is able to eliminate multiple impacts, so there isn't any bouncing back of the impactor resulting from elastic deformation of the specimen. The energy needed to create damage can be neglected for impacts that produce only small amounts of damage, so the energy-balance equation (1) can be written as:

$$\frac{1}{2}mv^2 = E_b + E_s + E_m + E_c \tag{1}$$

where m and v are the mass and the impact velocity of the impactor and the subscript b, s, m and c refer to energy dissipation in bending, shear, membrane and contact effects, respectively.

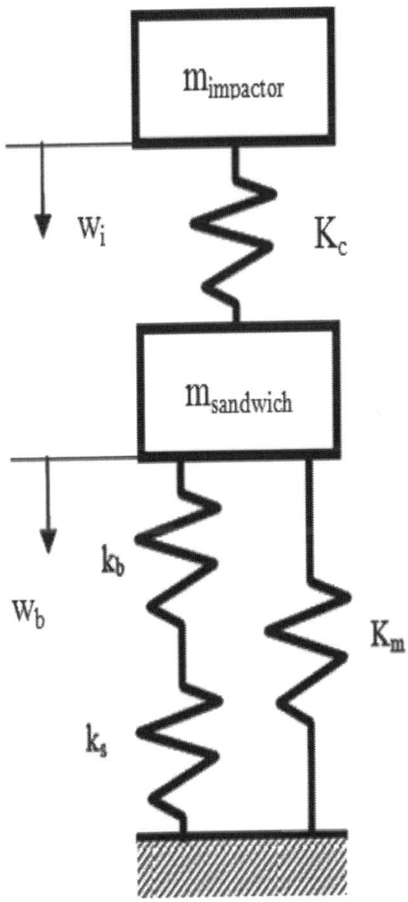

Figure 13. Spring-mass model for impact dynamics.

As the plate is thick, it can be assumed that the contribution of membrane forces can be neglected [14] as it is generally assumed in literature [1,6,14] for the energy balance model of the impact response of the sandwich panels.

The energy absorbed in bending and shear effects at maximum displacement (force) w_{max} (F_{max}) is equal to:

$$E_b + E_s = \frac{F_{max}^2}{2K_{bs}} \tag{2}$$

where K_{bs} is the linear stiffness including bending and shear effects.

Contact effects between the impactor and the sandwich structure can be calculated using the Meyer contact law, which relates the indentation depth α to the applied load F and is expressed by:

$$F = K_i \alpha^n \tag{3}$$

where K_i and n are constants that can be determined by fitting the experimental data of the indentation tests. The mathematical expression of the energy absorption due to contact effects is obtained by integrating Equation (3) between 0 and the maximum indentation value α_{max}:

$$E_c = \int_0^{\alpha_{max}} F d\alpha = \frac{F_{max}^{1+\frac{1}{n}}}{(n+1)K_i^{\frac{1}{n}}} \tag{4}$$

Finally, the energy balance for the sandwich structure can be obtained considering that the initial kinetic energy is equal to the energy absorption in bending, shear and contact effects:

$$\frac{1}{2}mv^2 = E_b + E_s + E_c = \frac{1}{2K_{bs}}F_{max}^2 + \frac{1}{(n+1)K_i^{\frac{1}{n}}}F_{max}^{1+\frac{1}{n}} \qquad (5)$$

This theoretical approach has been applied in the current study to investigate the impact response of *AFS* panels. This energy-balance model, once validated using data from tests carried out at some impact velocities, can be used to predict the maximum impact force for a given impact velocity or energy applying Equation (5).

The parameters of the energy balance model are generally determined in literature from the results of static tests. In this scientific research the parameters were obtained directly from the measurements carried out on the *CT* images (Figure 14); the vertical displacement w_b of the core at bottom face sheet interface and the impactor displacement w_i were measured by analysing the *CT* images of midplanes of the panels impacted at velocities lower than 5 ms^{-1}, that didn't produce the complete failure of the panels. The core compression displacement α is obtained by the subtraction between the impactor displacement w_i and the vertical displacement w_b of the core.

According to the spring-mass model, it can be assumed a linear relationship between the impact load and the corresponding global displacement w_b of the sandwich panel and the K_{bs} stiffness is the slope of this linear function. Moreover, the K_{bs} stiffness does not change with the impact velocity, so the K_{bs} stiffness was assessed in the present research paper by the slope of the linear function interpolating the peak loads at different impact velocities, obtained by the impact tests, versus the corresponding bending and shear deflections of the midplane of the panel, that were measured using the *CT*. The values of the force contact loads F_{max}, obtained by the experimental tests, were plotted versus the corresponding values of the deflection w_b of the panel midplane and a linear regression was performed in order to obtain the values of the stiffness K_{bs} for each *AFS* typology, as shown in Figure 15.

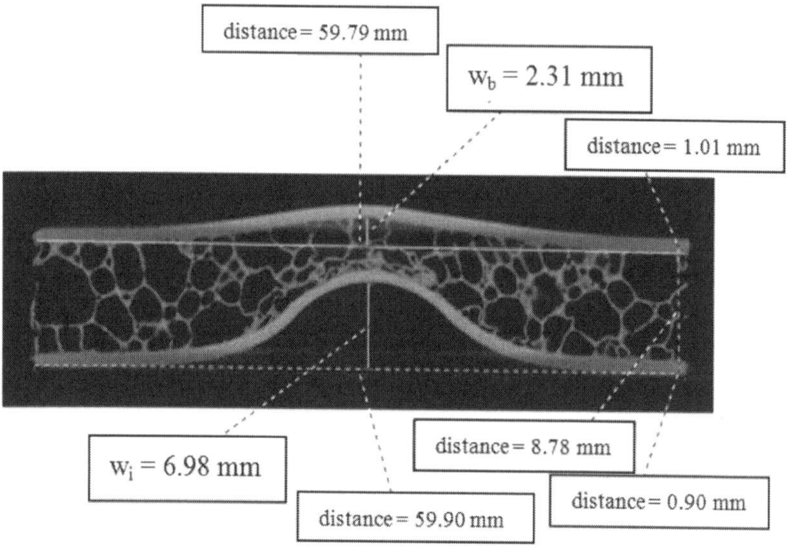

Figure 14. Measurements of the displacements w_i and w_b carried out on the *CT* image.

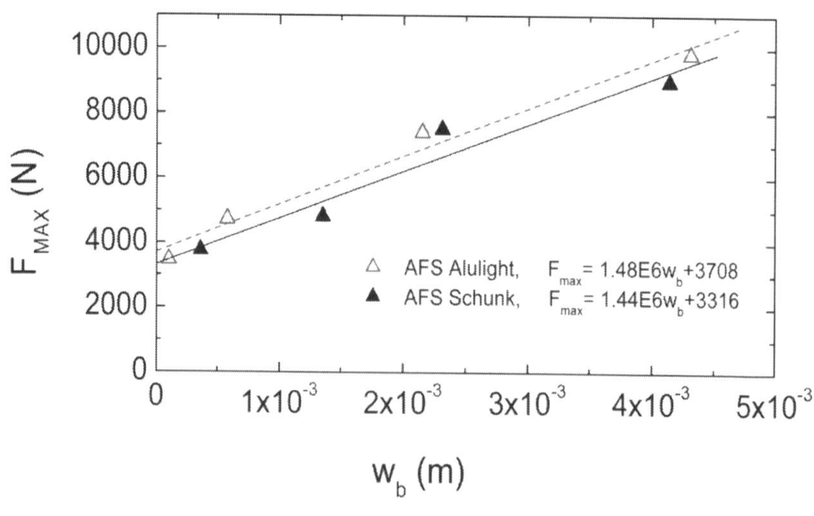

Figure 15. Peak load *vs.* sandwich deflection.

The contribution of the energy dissipation due to bending and shear E_{bs} was evaluated applying Equation (2) and the contact energy E_c was obtained simply by subtracting these values of E_{bs} from the values of the total dissipated energy.

Considering the bilogarithmic expression of Equation (4), a linear equation is obtained:

$$\log(E_c)=\left(1+\frac{1}{n}\right)\log(F_{max})-\left[\log(n+1)+\frac{1}{n}\log(K_i)\right] \tag{6}$$

The values of the contact energy E_c were plotted versus the corresponding values of the peak load F_{max} in a bilogarithmic scale and a linear regression was performed in order to assess the contact parameters (n, K_i) of equation (6), as shown inFigure 16; the value of n was obtained by the slope of the linear equation interpolating the data and the value of K_i by the constant term of the linear equation.

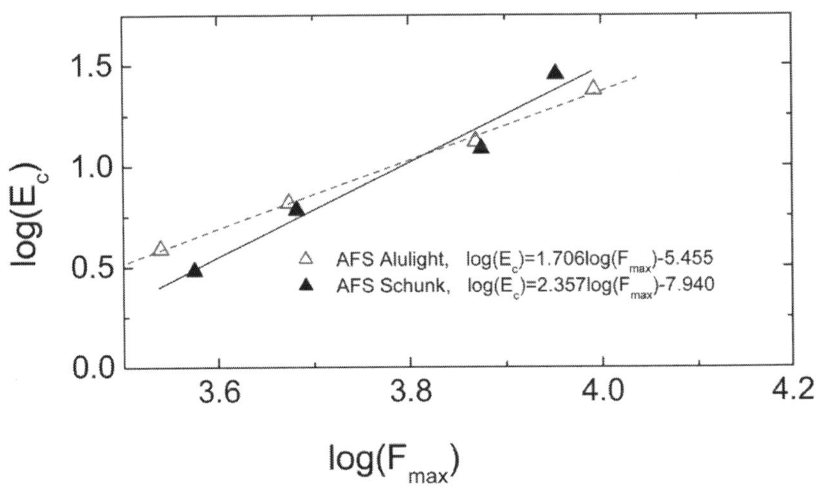

Figure 16. Contact energy dissipation *vs.* peak load.

The obtained values of the K_i and n constants are: $n =$ 1.42 and $K_i =$ 1.51E7 $Nm^{-1.42}$ for *AFS Alulight*, $n =$ 0.74 and $K_i =$ 4.70E5 $Nm^{-0.74}$ for *AFS Schunk*.

4. CONCLUSIONS

The impact behavior of aluminum foam sandwiches have been investigated by experimental tests and analytical approach, based on the energy balance model. The model parameters were obtained from the measurements carried out on the *CT* images of the impacted panels and not from the results of static tests, as it is usually done in literature.

The failure mode and the damage have been investigated by Computed Tomography. The investigation of the aluminum foams by means of the *3D CT* is very useful to check their quality in terms of porosity distribution, which influences the mechanical properties of the foams.

Low velocity impact tests on *AFS* structures have shown that the dynamic response of these sandwiches depends on the quality and mechanical properties of the foam core material.

The *AFS* structures are relatively intact compared to the more catastrophic and localized fracture of the polymeric sandwiches, so they are a better choice with respect to the polymeric sandwiches to realize progressive energy absorbers.

The obtained results have particular importance for applications that require lightweight structures with a high capacity of energy dissipation, such as the transport industry, where problems of collision and crash have increased in the last years. The use of these sandwich structures can lead to a weight reduction of the ships, providing an adequate structural strength under operating conditions.

REFERENCES

1. Abrate, S. *Impact on Composite Structures*; Cambridge University Press: Cambridge, UK, 2005.

2. Ashby, M.F.; Evans, A.; Fleck, N.A.; Gibson, L.J.; Hutchinson, J.W.; Wadley, H.N.G. *Metal Foams: A Design Guide*; Butterworth Heinmann: Burlington, VT, USA, 2000.

3. Gibson, L.J.; Ashby, M.F. *Cellular Solids*, 2nd ed.; Cambridge University Press: Cambridge, UK, 1997.

4. Banhart, J.; Schmoll, C.; Neumann, U. Light-weight aluminum foam structures for ships. Proceedings of the Conference on Materials in Oceanic Environment (Euromat '98), Lisbon, Portugal, 22–24 July 1998; Faria, L., Ed.; Federation of European Materials Societies (FEMS): Lisbon, Portugal, 1998; 1, pp. 55–63.

5. Baumeister, J.; Banhart, J.; Weber, M. Aluminum foams for transport industry. *Mater. Des.* **1997**, *18*, 217–220.

6. Hazizan, M.A.; Cantwell, W.J. The low velocity impact response of foam-based sandwich structures. *Compos. Part B***2002**, *33*, 193–204.

7. Crupi, V.; Epasto, G.; Guglielmino, E. Low velocity impact strength of sandwich materials. *J. Sandw. Struct. Mater.***2011**, *13*, 409–426.

8. Crupi, V.; Epasto, G.; Guglielmino, E. Computed tomography analysis of damage in composites subjected to impact loading. *Fract. Struct. Integr.* **2011**, *17*, 32–41.

9. Olurin, O.B.; Arnold, M.; Körner, C.; Singer, R.F. The investigation of morphometric parameters of aluminum foams using micro-computed tomography. *Mater. Sci. Eng.* **2002**, *A328*, 334–343.

10. Wicklein, M.; Thoma, K. Numerical investigations of the elastic and plastic behaviour of an open-cell aluminum foam.*Mater. Sci. Eng.* **2005**, *A397*, 391–399.

11. Mahfuz, H.; Al Mamum, W.; Jeelani, S. Effect of core density and implanted delamination on the high strain rate response of foam core sandwich composites. *Sandw. Constr.* **1997**, *5*, 597–606.

12. Compston, P.; Styles, M.; Kalyanasundaram, S. Low energy impact damage modes in aluminum foam and polymer foam sandwich structures. *J. Sandw. Struct. Mater.* **2006**, *8*, 365–379.

13. Shivakumar, K.N.; Elber, W.; Illg, W. Prediction of impact force and duration during low-velocity impact on circular composite laminates. *Trans. ASME J. Appl. Mech.* **1985**, *52*, 674–680.

14. Foo, C.C.; Seah, L.K.; Chai, G.B. Low-velocity impact failure of aluminum honeycomb sandwich panels. *Compos. Struct.* **2008**, *85*, 20–28.

CHAPTER 6

The Stiffness and Damping Characteristics of a Dual-Chamber Air Spring Device Applied to Motion Suppression of Marine Structures

Xiaohui Zeng [1,*], Liang Zhang [1], Yang Yu [2], Min Shi [1] and Jifu Zhou [1]

[1]Institute of Mechanics, Chinese Academy of Sciences, Beijing 100190, China

[2]Collaborative Innovation Center for Advanced Ship and Deep-Sea Exploration, State Key Laboratory of Hydraulic Engineering Simulation and Safety, Tianjin University, Tianjin 300072, China

ABSTRACT

Dual-chamber air springs are used as a key component for vibration isolation in some industrial applications. The working principle of the dual-chamber air spring device as applied to motion suppression of marine structures is similar to that of the traditional air spring, but they differ in their specific characteristics. The stiffness and damping of the dual-chamber air spring device determine the extent of motion suppression. In this article, we investigate the stiffness and damping characteristics of a dual-chamber air spring device applied to marine structure motion suppression using orthogonal analysis and an experimental method. We measure the effects of volume ratio, orifice ratio, excitation amplitude, and frequency on the stiffness and damping of the dual-chamber vibration absorber. Based on the experimental results, a higher-order non-linear regression method is obtained. We achieve a rapid calculation model for dual-chamber air spring stiffness and damping, which can provide guidance to project design.

KEYWORDS

stiffness and damping; orthogonal analysis; vibration suppression; floating marine structure; dual-chamber air spring

1. INTRODUCTION

Marine structure safety, especially in the deep sea, is an important concern. The deep-sea floating platform is one of the most important large structures used in ocean energy exploitation. Dynamic responses due to the wind and wave effects must be considered to ensure a safe design [1]. The combined action of complex multiple loads could cause a large vibration amplitude in marine structures [2,3]. Large amplitude and alternating vibration are the main detrimental effects that reduce the safety and fatigue life of a deep-sea platform. In the field of civil engineering, engineers use a wide variety of energy-absorbing devices to reduce the vibration of building structures. These energy-absorbing devices may also be applied to the problem of vibration suppression of deep-sea marine structures.

The vibration of a floating platform could be reduced using active or passive controls [4,5] and various devices have been proposed. A Tuned Liquid Column Damper (TLCD), a kind of passive vibration control device to suppress movement, was proposed by Sakai [6] in 1989. Lee *et al.* [7,8] first used the TLCD to suppress the surge and sway motions vibration of a Tension Leg Platform (TLP). Taflanidis *et al.* [9] further developed a simulation-based method for the design of mass dampers applied for the response mitigation of tension lag platforms. Lee *et al.* [10] experimentally studied the harmonic responses of the TLCD for wind excitations. Tanmoy *et al.* [11] studied the effective performance of TLCD and another passive vibration-mitigating device, a tuned liquid column ball damper for the control of wave-induced vibration. Zeng *et al.* [12] invented a new type of energy-absorbing device, the S-shaped TLCD. The S-shaped TLCD could effectively suppress the horizontal movement and vertical in-plane rotation of a TLP. Although these devices were effective for the suppression of surge and sway motion, TLCD was ineffective at the suppression of vertical movement, and sometime even enhanced this movement [10]. The vertical movement of the floating structures significantly impacts the strength and fatigue of the tension leg and mooring system, and the inability of these devices to suppress this movement could decrease the safety and service life of the structure.

Vibration-induced vertical motion of TLPs could be balanced by using absorbers with air springs. Dual-chamber air springs are an effective energy-absorbing device used as a key component for vibration isolation. Rijken, Bian, and Spillane*et al.* [13,14,15] applied a system of vibration absorbers using dual-chamber air springs and water columns to suppress resonant motions and studied the effects of the orifice ratio for structural damping. Bachrach and Rivin [16] studied the complex dynamic stiffness of the damper spring, a function of the excitation frequency. The experimental results measured by Kim and Lee [17] for a dual-chamber pneumatic spring exhibited significant amplitude-dependent nonlinear behavior. Jing *et al.* [18,19,20] proposed a characteristic output spectrum nonlinear (nCOS) method. The nCOS method is based on the theory of Volterra series expansion. They used the nCOS method for analysis and design of an air spring in a nonlinear vehicle suspension system. These studies focused on traditional forms of a dual-chamber air spring.

A dual-chamber air spring device applied to marine structure motion suppression requires a gas-liquid coupling air spring structure. This differs from the traditional air spring both in structure type and in the vibration characteristics. The compression of the gas in the traditional air spring was controlled by a piston, which acted on the pressure surface of the air spring [21,22,23]. As shown in Figure 1a, the displacement of the piston, x_0, causes a compression of the air in the upper chamber. The pressure variation of the air spring chamber is of the same phase and amplitude as the external load. For the gas-liquid coupling air spring device applied to motion suppression of marine structures, a liquid column can also move in an independent manner. As shown in Figure 1b, the amount of compressed air in the air chamber is the relative displacement between the liquid column and the top of the air spring, *i.e.* $x_1 - x_0$. The movement of the liquid column (similar to the piston) and the external excitation can differ in phases and amplitude. The stiffness and damping characteristics of this kind of dual-chamber air spring have not been investigated.

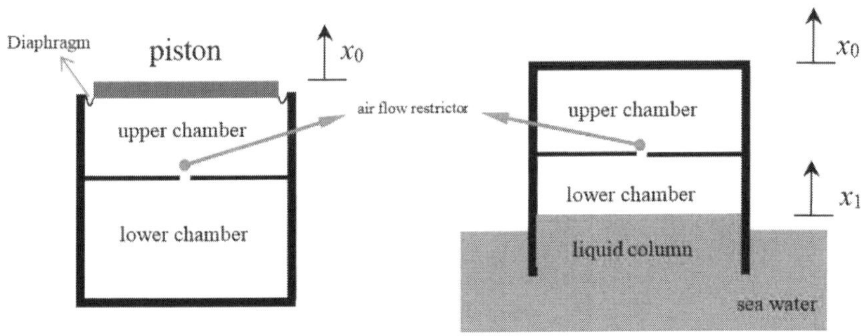

Figure 1. Schematic diagram of two kinds of dual-chamber air spring: (**a**) the traditional device; (**b**) a gas-liquid coupling device.

The two main determinants of structure movement suppression effect are the stiffness and damping of the energy-absorbing device. Previous studies of damping and stiffness focused on the traditional air spring, and used an approximation to simulate air flow through the orifice plates [24,25,26]. These results may not apply to the gas-liquid coupling dual-chamber air spring device. Moreover, few factors were investigated and analyzed. There is not a comprehensive study on which factors affect the stiffness and damping characteristics of dual-chamber air spring. Further studies must be carried out on the gas-liquid coupling dual-chamber air spring device.

Here, we investigate the damping and stiffness of a dual-chamber air spring device applied to motion suppression of marine structures. We used orthogonal analysis and an experimental method and found that stiffness and damping are a function of the volume ratio, orifice ratio, excitation amplitude, and frequency. We then measured the effects of these four factors on dual-chamber vibration absorber stiffness and damping. Based on the experimental results, we used higher-order non-linear regression method and generated a rapid calculation model for dual-chamber air spring stiffness and damping, which can provide guidance for future project design.

2. THEORETICAL ANALYSIS

The gas-liquid coupling, dual-chamber, air spring energy-absorbing device has two chambers. The upper chamber functions for air storage and has a fixed volume, while the volume of the lower chamber, in contact with the water, varies with the motion of the water column. The gas and the oscillating liquid column are coupled and this device uses the interaction of oscillations of the liquid column and gas to achieve floating structure motion suppression.

The performance and design of the dual-chamber air spring device are usually modeled after a spring damper system. The characteristics of the damping device are described by the spring stiffness and damping coefficient [26,27,28,29,30]. In this work, the dual-chamber air spring device is a complex fluid-gas coupling system, but its dynamic behavior can also be characterized by the spring stiffness and damping coefficient. The physical model of the common spring damping system can be described using a parallel model or as a series model. Here, we explore the advantages and disadvantages of the two models to characterize the dynamic characteristics of the dual-chamber air spring device applied to motion suppression of marine structures.

In this section, three kinds of energy dissipation expression are derived: a series model, a parallel model, and work done by excitation force.

2.1. Parallel Model

As shown in Figure 2, the standard equation of the model is:

$$M\ddot{X} + C\dot{X} + KX = F \tag{1}$$

where *M* represents the overall mass of the vibration absorber structure, *C* represents the overall damping of the structure, and *K* refers to the overall stiffness of the structure.

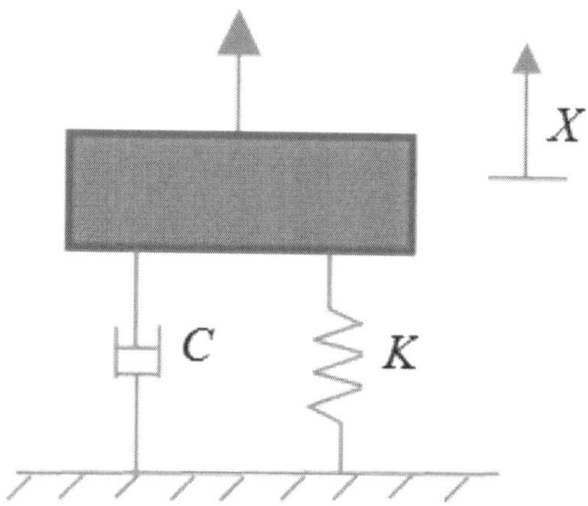

Figure 2. Parallel model.

The structure stiffness in the above equation changes with the air pressure in the vibration absorber air chamber. The change in air pressure is caused by the relative motion of the liquid column in the air chamber. The damping has two components. Some damping is due to fluid vortex shedding at the bottom of the vibration absorber and the friction between the outer wall and water and this damping is related to the motion of the vibration absorber structure. Damping is also caused by the effect of the air in the air chamber and the orifice as well as the friction between inner wall and water, and this damping is related to the motion of the liquid column in the vibration absorber to the vibration absorber.

For harmonic excitation F=F0eiωt, the structural response *X* is also a harmonic motion which can be represented as X=X0ei(ωt−φ), substituted into Equation (1) and using Euler's Formula, eiφ=cosφ+isinφ, the stiffness and damping of the vibration system can be written as:

$$K = \frac{F_0}{X_0}\cos\varphi + \omega^2 M \quad (2)$$

$$C = \frac{1}{\omega}\frac{F_0}{X_0}\sin\varphi \qquad (3)$$

The mass matrix can be M=[m1,m2,m3,m4,m5,m6,m7]T. In the experiment, the mass of each different air chamber length structure is shown as follows:

m1=5.769,m2=5.967,m3=6.178,m4=6.386,m5=6.593,m6=6.773,m7=7.001, in kg.

The periodical energy consumption of damping is:

$$Q = \int C\dot{X}dX = \int_0^T \left(C\dot{X}\right)\dot{X}dt = \int_0^T C\dot{X}^2 dt = C\int_0^T \dot{X}^2 dt = \frac{1}{\omega}\frac{F_0}{X_0}\sin\varphi \sum_{i=1}^{i=f_s\cdot T} \dot{X}_i^2 t_i \qquad (4)$$

where f_s refers to the sampling frequency of the experiment, $f_s = 1024$ Hz. Due to the equal-interval sampling of the experiment data, $t_i = \frac{1}{1024}$.

2.2. Series Model

As shown in Figure 3, the standard equation of the model is:

$$\begin{cases} M\ddot{X} + k(X - x) = F \\ k(x - X) + c\dot{x} = 0 \end{cases} \qquad (5)$$

Substitute F=F0eiωt and X=X0ei(ωt−φ) into the equation and apply Euler's Formula, eiφ=cosφ+isinφ, to determine the damping and stiffness:

$$c = \frac{F_0}{\omega X_0}\sin\varphi + \frac{F_0}{\omega X_0}\cot\varphi\cos\varphi + 2\omega M\cot\varphi + \frac{\omega^3 M^2 X_0}{F_0\sin\varphi} \tag{6}$$

$$k = \frac{c}{\frac{1}{\omega}\cot\varphi + \frac{\omega M X_0}{F_0\sin\varphi}} \tag{7}$$

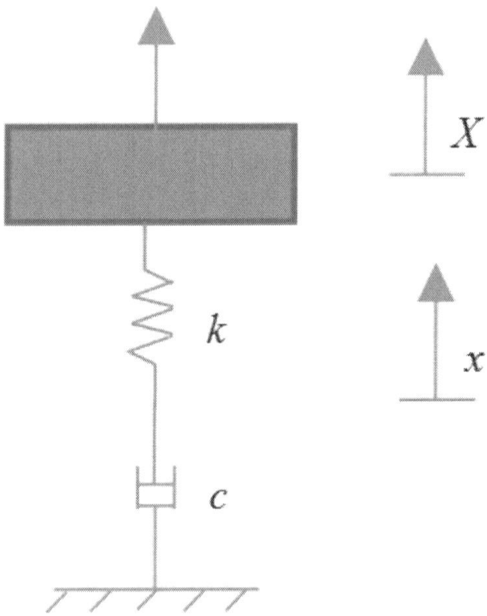

Figure 3. Series model.

The periodical energy consumption of damping is:

$$q = \int c\dot{x}dx = \int_0^T c\dot{x}^2 dt = \sum_{i=1}^{i=f_s \cdot T} c\dot{x}_i^2 t_i \tag{8}$$

2.3. Work Done by Excitation Force

The periodical work done by excitation force of the system is:

$$W = \int F dX = \int_{0}^{T} F\dot{X} dt = \sum_{i=1}^{i=f_s \cdot T} F_i \dot{X}_i t_i \tag{9}$$

where F refers to the excitation load, and X refers to the structural response.

3. EXPERIMENTAL METHOD

3.1. Design of Experimental Device

The dual-chamber air spring vibration absorber described here is mainly used to restrain the large vibration of an offshore structure. In the experiment, a large water container was designed to simulate the action of sea water. The container is a cylindrical device with a diameter of 1500 mm (much larger than the diameter of 100 mm of the air chamber). As shown inFigure 4, the experimental device consists of: 1. Hydraulic pressure loader; 2. Force sensor; 3. Dual-chamber air spring; 4. Orifice plate; and 5. Air pressure sensor.

As shown in Figure 5, in order to study the effects of different upper and lower air chamber ratios on the vibrating characteristics, the length of the lower air chamber (200 mm) was kept constant while the upper air chamber was changed（100 mm, 150 mm, 200 mm, 250 mm, 300 mm, 350 mm, 400 mm） to study the effect of different volume ratios. The upper and lower air chambers are connected by an orifice plate and the air flows back and forth through an orifice. The orifice changes the air distribution of the chambers greatly and can influence the vibrating characteristics of the structure. In this experiment, the orifice diameters used to study the effect of different orifice ratios on the air spring were 0, 10, 20, 30, 40, 50, and 60 mm.

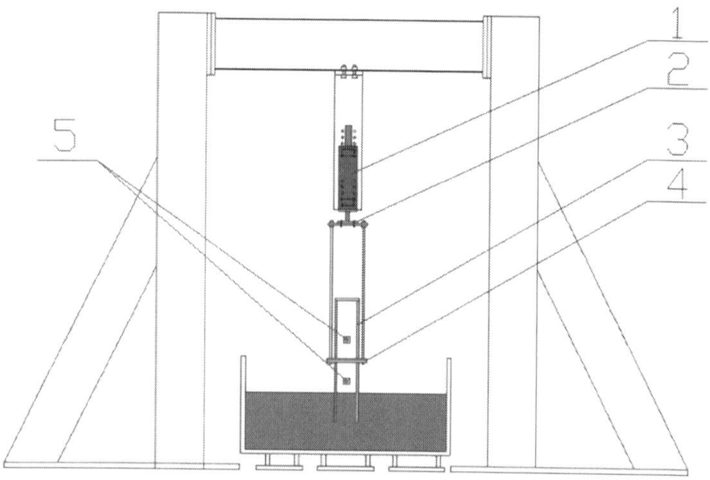

Figure 4. Schematic diagram for the experimental device.

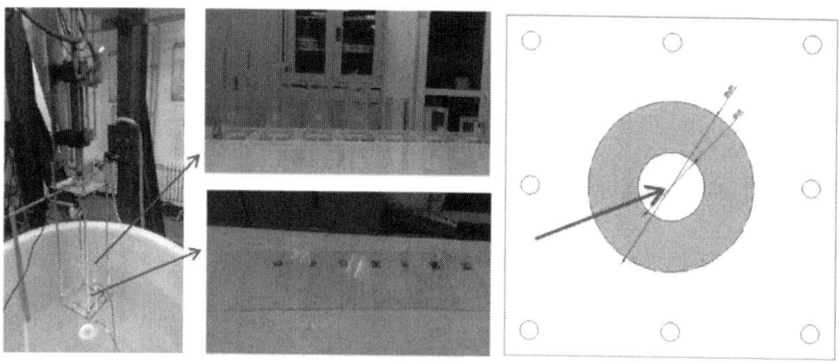

Figure 5. Upper chamber and orifice plates of the dual-chamber air spring.

As shown in Figure 6, a hydraulic servo fatigue machine is used as the loading system of the experiment. The fatigue machine has excellent performance and can be used for large structure experiments. By controlling the loading frequency and amplitude, we can test the effect of external load on the vibrating characteristics of the air spring. In the experiment, the amplitude range was 2, 4, 6, 8, 10, 12, and 14 mm, and the frequency range was 0.497–5.474 Hz.

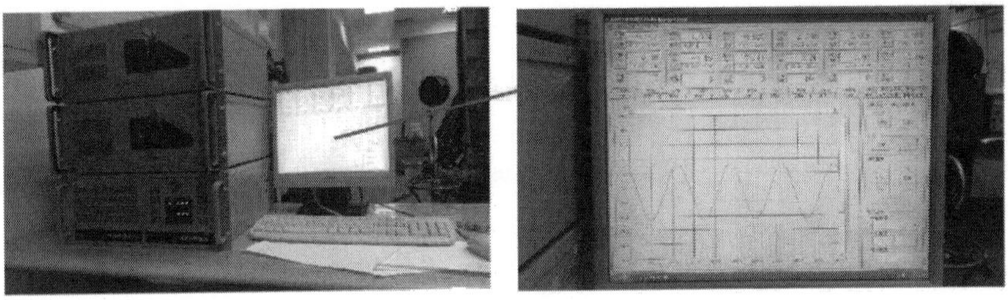

Figure 6. Loading system for experimental device.

3.2. Setting of Test Parameters

We studied the influences of each factor on dual-chamber air spring dynamic stiffness and damping. The experiment assumes the cross section of the upper air chamber equals that of the lower air chamber.

Variables to be considered: Volume of upper air chamber V_2, Volume of lower air chamber V_1, opening diameter d, amplitude of outer excitation A, and frequency of outer excitation ω.

The length of the lower air chamber, L_1, and the frequency of outer excitation, ω, were used as the basic variables with the following three dimensionless numbers:

Height of lower air chamber $L_1 = 0.2$ m

Diameter of air chamber $D = 0.1$ m

$\xi_1 = \frac{L_2}{L_1}$, volume ratio, values = 0.5, 0.75, 1, 1.25, 1.5, 1.75, 2

$\xi_2 = \frac{d}{D}$, orifice ratio, values = 0, 0.1, 0.2, 0.3, 0.4, 0.5, 0.6

$\xi_3 = \frac{A}{L_1}$, ratio of the amplitude of outer excitation and the height of air chamber, values = 0.01, 0.02, 0.03, 0.04, 0.05, 0.06, 0.07.

Frequency of outer excitation. Log value $\ln\omega$ was used to investigate the vibrating characteristics of the air spring at low frequency (values = −0.7, −0.3, 0.1, 0.5, 0.9, 1.3, 1.7).

In total, four variables and seven values were tested, as shown in Table 1:

Table 1. Four factors and seven levels chart.

Level / Factor	Volume Ratio (ξ_1)	Orifice Ratio (ξ_2)	Loading Amplitude Ratio (ξ_3)	Loading Frequency $\ln\omega$	Real Loading Frequency (Hz)
1	0.5	0.0	0.01	−0.7	0.497
2	0.75	0.1	0.02	−0.3	0.741
3	1.0	0.2	0.03	0.1	1.105
4	1.25	0.3	0.04	0.5	1.649
5	1.5	0.4	0.05	0.9	2.460
6	1.75	0.5	0.06	1.3	3.669
7	2.0	0.6	0.07	1.7	5.474

3.3. Orthogonal Experiment and Orthogonal Array

For experimental designs with multiple factors and multiple levels, a large number of experiments are required. For example, for the experimental scheme with four variables and seven levels in this paper, 2401 (7^4) experiments need to be carried out. Hence, a method with the minimum numbers of experiments is needed to find out the effect of each parameter on the vibrating characteristics of the air spring.

In 1947, Rao [31] developed factorial experiments, an optimization experimental method (an efficient testing strategy) that can determine the relation between factors and a test index and specify the primary and secondary factors using fewer trials. Taguchi [32] effectively used orthogonal arrays in his research. Azouzi and Guillot [33] examined the feasibility for an intelligent sensor fusion to estimate online surface finish and dimensional deviations using orthogonal arrays. Green, Krieger, and Wind [34] applied orthogonal arrays to conjoint analysis. The use of the orthogonal test method minimizes the number of tests but still allows determination of the changing rules of all factors. We applied the orthogonal experiment method to investigate the effect of upper and lower air chamber volume ratio, orifice ratio, loading amplitude, and loading frequency of outer excitation.

As an effective method of solving multi-factor experiment problems, orthogonal experimental design selects typical points from the overall experiment and tests them. These typical points are "even" and "regular": (1) in the experiment, each factor has the same occurrence number at each different level; (2) each combination of two factors at each different level occurred in the experiment and had the same occurrence number. Meeting these two criteria indicates that the experiment scheme designed by orthogonal experiment method is typical. This method reduces the number of experiments and can generally reflect the overall effect of each factor at each level on the index.

Range analysis is commonly used in scientific experiments. Change in experimental results often occurs in scientific experiments due to two types of factors. One is variations arising from random effects; such effects are controllable in experiments, and are therefore inevitable. The other is artificial control produces changes in experimental results. When such factors have a significant impact on the experiment, they are bound to significantly alter the results, accompanied by random factors. Conversely, when such factors have no significant effect on the experimental results, corresponding changes will not be manifested obviously, so changes in experimental results have been substantially ascribed to the effects of random factors. The purpose of conducting scientific experiments is often to determine whether these artificially controllable factors have an impact on the experimental results and what the effect is.

Range analysis is an effective tool to judge the above matters through analyzing the data variation in experimental results, because it can separate random variation from non-random variation in a hybrid state to help determine the source of dominant variations. Using range analysis, we analyzed the results of orthogonal design to explore the effect of various factors on air spring characteristics. In addition, we investigated the effects of four factors (upper and lower air chamber volume ratio, orifice ratio, loading amplitude, and loading frequency of outer excitation) on the air vibration characteristics of the dual-chamber. Seven variation levels of each factor were examined. The tests were performed 49 times, accounting for only 2% of the traditional number of trials. The arrangements and results of the orthogonal analysis scheme are shown inAppendix 1.

4. EXPERIMENTAL RESULTS AND ANALYSIS OF A DUAL-CHAMBER AIR SPRING

We used an intuitive method in orthogonal analysis that looks at means and ranges. The mean average is the average value of different factors at the same level, and this can reveal the effect of different levels of single factors on the indicators. The range indicates the maximum value of the average numerical difference for each factor at each level, and reflects the impact of each column factor at different levels, from which the primary and secondary sequence of factors can be determined. If each factor has the same number of level, the influence extent of each factor can be judged directly by comparing the range size.

For 49 sets of experimental schemes, the comparison of energy dissipation is shown in Figure 7.

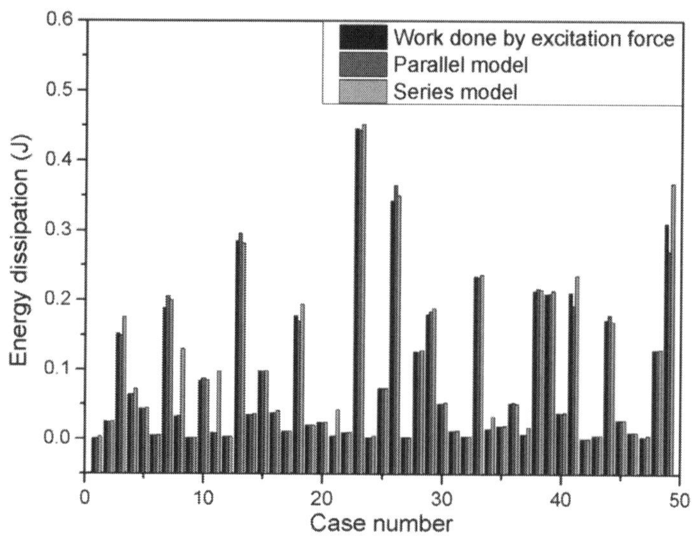

Figure 7. Energy dissipation test program comparison chart.

Figure 7 shows that under most operating conditions, there is a good fit between the parallel model and work done by excitation force, but under individual conditions, the series model shows better goodness of fit. From the relationship of energy dissipation mean of the 49 sets of orthogonal design schemes, the goodness of fit of the parallel model $\alpha_p = 1 - \frac{\sum_{0}^{49} \frac{|Q_{parallel} - Q_{structure}|}{Q_{structure}}}{49} = 99.88\%$.

The goodness of fit of the series model $\alpha_s = 1 - \frac{\sum_{0}^{49} \frac{|Q_{series} - Q_{structure}|}{Q_{structure}}}{49} = 89.43\%$.

By the range relation of energy consumption, we can infer a higher goodness of fit between work done by excitation force and energy dissipation of the parallel model. Furthermore, the parallel model can be employed to make deeper analyses of the structure.

4.1. Influence of Various Parameters on Air Spring Stiffness

Stiffness is a function of the volume ratio, orifice ratio, excitation amplitude, and frequency. The influence law can be seen from the following figures:

As can be seen from Figure 8, the effect of the volume ratio on the dual-chamber vibration absorber stiffness is not monotonic. With an increase in the volume ratio, the stiffness of vibration absorber first slowly decreases and increases at a volume ratio of 1.25, then decreases and finally increases dramatically. As per Figure 9, the effect of the orifice ratio on the dual-chamber vibration absorber stiffness similarly is not monotonic. With an increase in the orifice ratio, the stiffness of vibration absorber first shows little change, then increases at a orifice ratio of 0.4, then decreases and finally increases to its highest level.

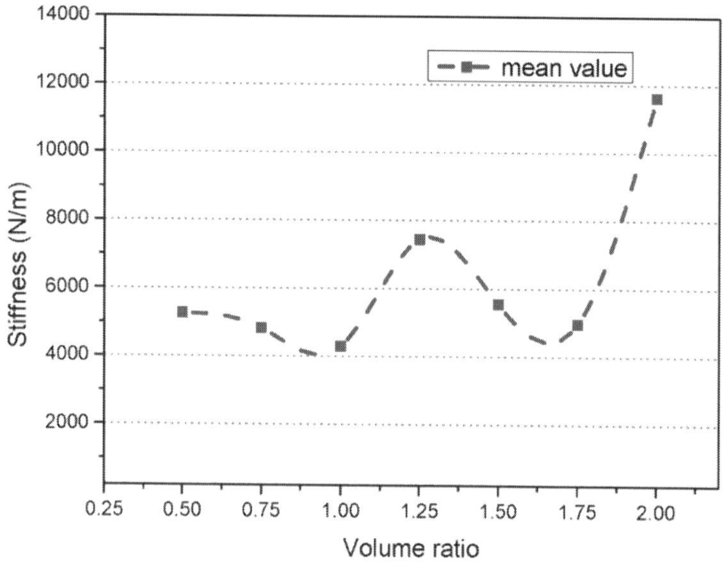

Figure 8. Effect of the volume ratio on the dual-chamber air spring stiffness.

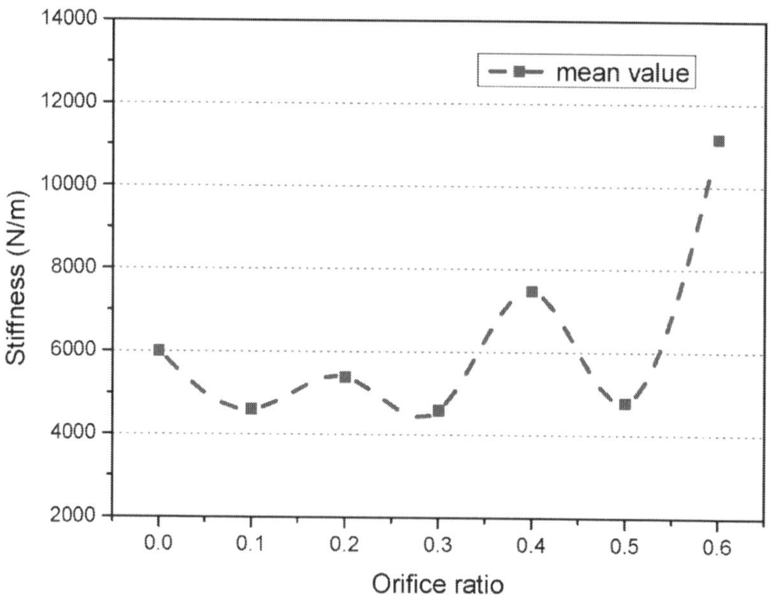

Figure 9. Effect of the orifice ratio on the dual-chamber air spring stiffness.

According to Figure 10, the effect of the loading amplitude ratio on the dual-chamber vibration absorber stiffness is basically monotonic. With an increase in the amplitude ratio, the stiffness of the dual-chamber air spring also increases and slightly protrudes at the amplitude ratio of 0.04. Figure 11 shows the effect of the loading frequency on the dual-chamber vibration absorber stiffness is also monotonic. The stiffness of the dual-chamber air spring monotonically increases with an increase in the loading frequency. As can be seen from Figure 12, the effect of each factor on the dual-chamber vibration absorber stiffness varies, where the volume ratio, orifice ratio, and loading amplitude ratio exert considerable influence but the loading frequency has the maximum impact. The results described in Figure 12 are only applicable when considering the volume ratio, the orifice ratio, load amplitude, and frequency in this case.

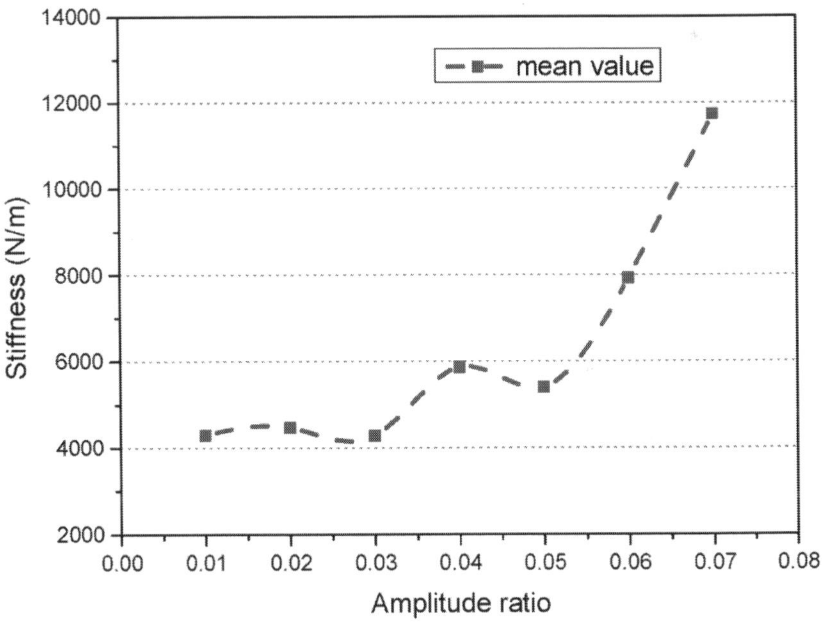

Figure 10. Effect of the amplitude ratio on the dual-chamber air spring stiffness.

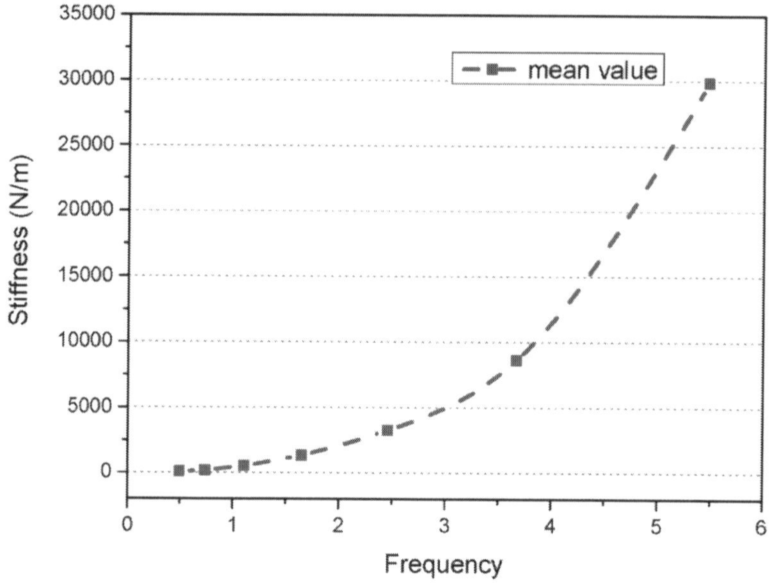

Figure 11. Effect of the frequency on the dual-chamber air spring stiffness.

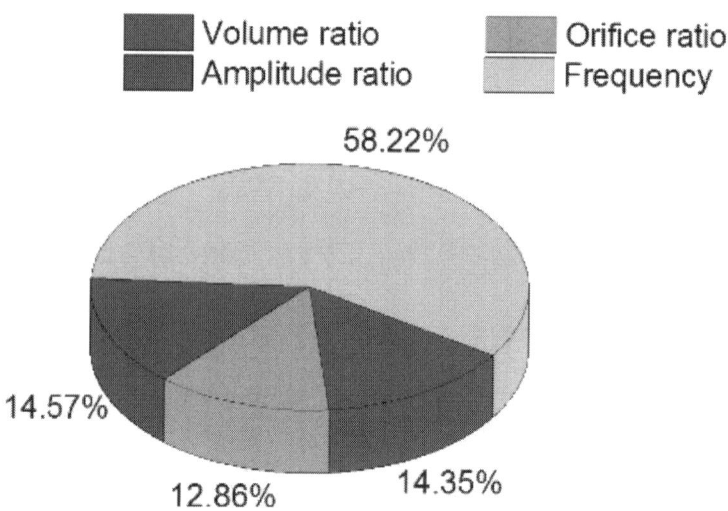

Figure 12. Summary of the effects of the four factors on the dual-chamber air spring stiffness.

4.2. Influence of Various Parameters on Air Spring Damping

Damping is a function of the volume ratio, orifice ratio, excitation amplitude, and frequency. In this section the four factors for the dual-chamber air spring damping effects are examined.

The influence law can be seen from the following figures:

Figure 13 shows that the effect of the volume ratio of the dual-chamber vibration absorber damping is not monotonic. With an increase in the volume ratio, the damping of the vibration absorber first slowly decreases and then increases at a volume ratio of 1.25, then decreases and then increases. As per Figure 14, the effect of the orifice ratio on the dual-chamber vibration absorber damping is also not monotonic. With an increase in the orifice ratio, the damping of vibration absorber first shows little change, then increases at an orifice ratio of 0.4, then decreases and then increases.

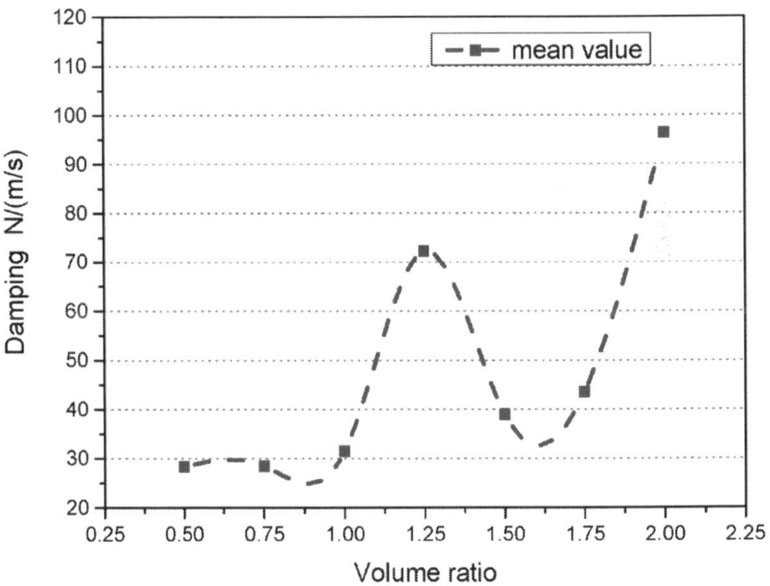

Figure 13. Effect of the volume ratio on the dual-chamber air spring damping.

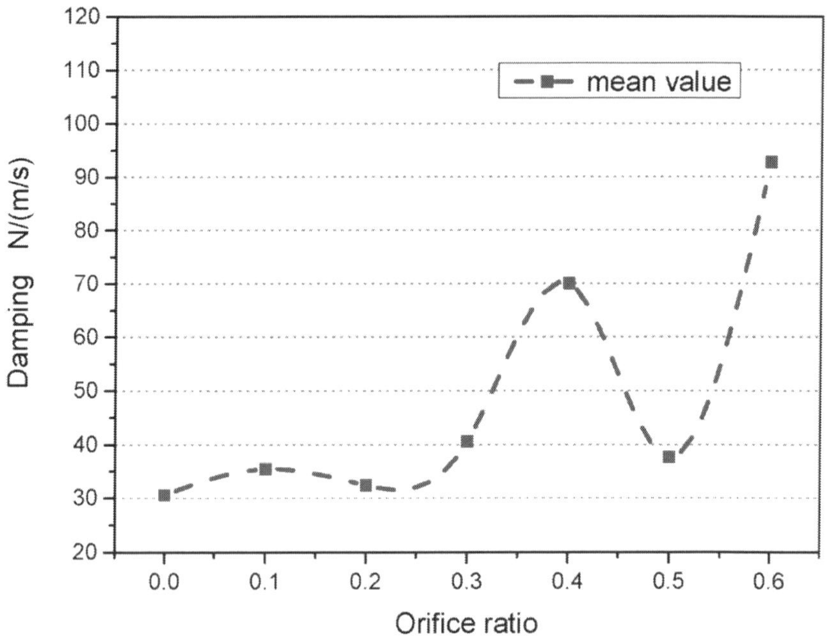

Figure 14. Effect of the orifice ratio on the dual-chamber air spring damping.

According to Figure 15, the effect of the loading amplitude ratio on the dual-chamber vibration absorber damping is not monotonic. With an increase in the amplitude ratio, the structural damping shows little change at first and then declines at 0.05, before eventually increasing. Figure 16 indicates that the effect of the loading frequency on the dual-chamber vibration absorber damping is basically monotonic. With an increase in the amplitude ratio, the vibration absorber damping increases until 3.669 Hz, and then increases dramatically. As summarized in Figure 17, the effect of each factor on the dual-chamber vibration absorber damping varies, where the volume ratio, orifice ratio, and loading amplitude ratio have strong effects but the loading frequency has the maximum impact. The results described in Figure 17 are only applicable when considering the volume ratio, the orifice ratio, load amplitude, and frequency in this case.

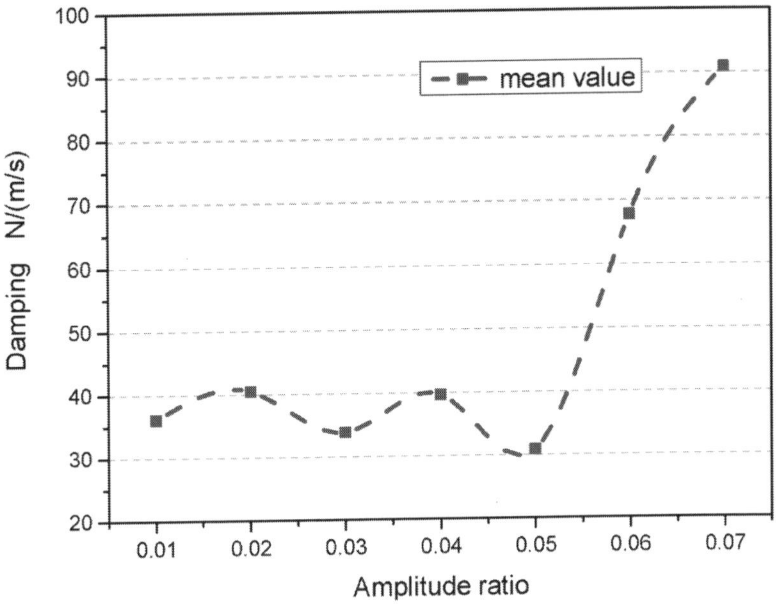

Figure 15. Effect of the amplitude ratio on the dual-chamber air spring damping.

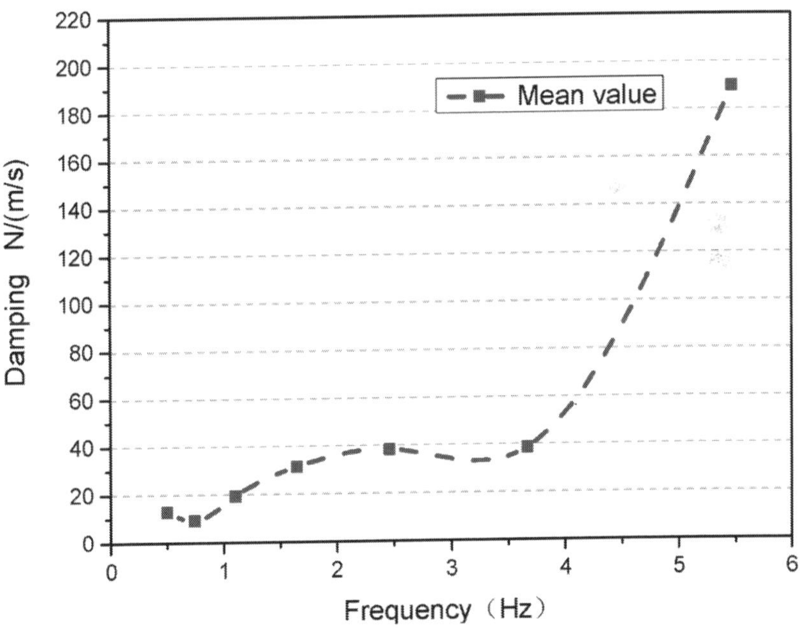

Figure 16. Effect of the frequency on the dual-chamber air spring damping.

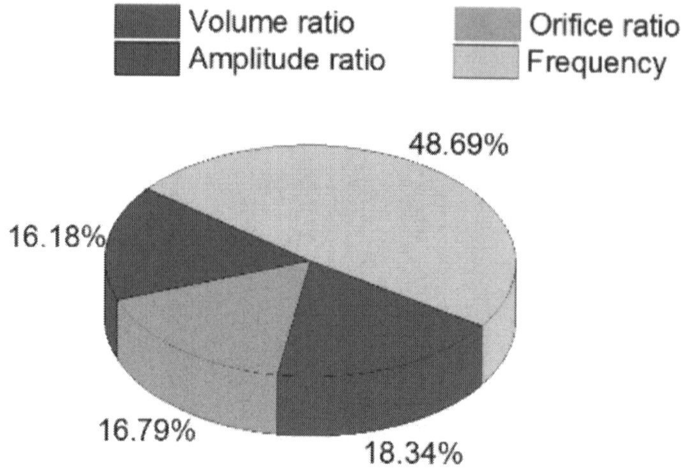

Figure 17. Summary of the effects of the four factors on dual-chamber air spring damping.

Thus, the four factors influence the stiffness and damping of the gas-liquid coupling dual-chamber air spring in a complex manner; there is not a simple linear relationship. Each factor has different effects on stiffness and damping. Prediction of the effects on stiffness and damping of the gas-liquid coupling dual-chamber air spring based on these four factors would be useful for efficient design, and in the next chapter, we propose an efficient model of dual-chamber air spring stiffness and damping to do this.

5. RAPID CALCULATION MODEL OF DUAL-CHAMBER AIR SPRING STIFFNESS AND DAMPING

From the previous analyses, we can see that volume ratio, orifice ratio, loading amplitude, and loading frequency all have complex effects on the stiffness and damping of the dual-chamber air spring device and are not monotonic in most cases. Therefore complex models must be constructed using statistical analyses of the experimental data.

5.1. Normalization Process of Each Factor

In order to facilitate the analysis of each factor's effect, the normalization process of each factor is performed.

Normalized volume ratio $\varphi_1 = \frac{\xi_1}{2}$ values of 0.25, 0.375, 0.5, 0.625, 0.75, 0.875, 1.

Normalized orifice ratio $\varphi_2 = \frac{\bar{\xi_2}}{0.6}$, values of 0, 0.167, 0.333, 0.5, 0.667, 0.833, 1.

Normalized loading amplitude ratio $\varphi_3 = \frac{\xi_3}{0.07}$, values of 0.142, 0.286, 0.429, 0.571, 0.714, 0.857, 1.

Normalized loading frequency $\varphi_4 = \frac{\ln\omega + 0.7}{2.4}$, values of 0, 0.167, 0.333, 0.5, 0.667, 0.833, 1.

The functional relationship between the stiffness and damping of the dual-chamber air spring and each factor is:

$$K_{\text{modle}} = f_1(\varphi_1, \varphi_2, \varphi_3, \varphi_4) \tag{10}$$

$$C_{\text{modle}} = f_2(\varphi_1, \varphi_2, \varphi_3, \varphi_4) \tag{11}$$

5.2. Determination of Rapid Calculation Model of Stiffness and Damping

From the above analysis, each factor has a nonlinear impact on the stiffness and damping of the dual-chamber air spring. The quadratic function can not accurately simulate the variation law of the stiffness and damping of the dual-chamber air spring. When the power of an independent variable is more than three, the regression function becomes very unstable [35], thus destabilizing the application of the regression model. In order to more accurately reflect the effect law of each factor on the stiffness and damping of the dual-chamber air spring, this paper uses cubic polynomials, which can usually be converted into ordinary multiple linear regression for processing. Experimental data are used to fit the expression of the stiffness and damping of the dual-chamber air spring.

Suppose the stiffness experimental value meets the following cubic polynomials:

$$
\begin{aligned}
f_1 &= \beta_0 + \sum_{j=1}^{34} \beta_j g_i(\varphi_1, \varphi_2, \varphi_3, \varphi_4) \\
&= \beta_0 + \beta_1\varphi_1 + \beta_2\varphi_2 + \beta_3\varphi_3 + \beta_4\varphi_4 + \beta_5\varphi_1^2 + \beta_6\varphi_2^2 + \beta_7\varphi_3^2 + \beta_8\varphi_4^2 + \beta_9\varphi_1\varphi_2 + \beta_{10}\varphi_1\varphi_3 + \beta_{11}\varphi_1\varphi_4 \\
&+ \beta_{12}\varphi_2\varphi_3 + \beta_{13}\varphi_2\varphi_4 + \beta_{14}\varphi_3\varphi_4 + \beta_{15}\varphi_1^3 + \beta_{16}\varphi_2^3 + \beta_{17}\varphi_3^3 + \beta_{18}\varphi_4^3 + \beta_{19}\varphi_1^2\varphi_2 + \beta_{20}\varphi_1^2\varphi_3 + \beta_{21}\varphi_1^2\varphi_4 \\
&+ \beta_{22}\varphi_1\varphi_2^2 + \beta_{23}\varphi_1\varphi_3^2 + \beta_{24}\varphi_1\varphi_4^2 + \beta_{25}\varphi_2^2\varphi_3 + \beta_{26}\varphi_2^2\varphi_4 + \beta_{27}\varphi_2\varphi_3^2 + \beta_{28}\varphi_2\varphi_4^2 + \beta_{29}\varphi_3^2\varphi_4 + \beta_{30}\varphi_3\varphi_4^2 \\
&+ \beta_{31}\varphi_1\varphi_2\varphi_3 + \beta_{32}\varphi_1\varphi_2\varphi_4 + \beta_{33}\varphi_1\varphi_3\varphi_4 + \beta_{34}\varphi_2\varphi_3\varphi_4 + \varepsilon_i(o^4)
\end{aligned}
\tag{12}
$$

where β0,β1,β2,....,β34 are undetermined coefficients; φ1,φ2,φ3,φ4 are controllable variables in the experiment; gj(φ1,φ2,φ3,φ4) is function expression corresponding to βj; and f1 is the stiffness of dual-chamber air spring.

This paper uses the least squares method to estimate the undetermined coefficients in the above formula to obtain the regression equation for stiffness:

$$
\begin{aligned}
\frac{\hat{f}_1}{E13} &= -0.0952 + 1.4139\varphi_1 - 0.6852\varphi_2 - 1.3314\varphi_3 + 1.2366\varphi_4 - 0.9296\varphi_1^2 + 0.5123\varphi_2^2 \\
&+ 0.3488\varphi_3^2 - 0.7531\varphi_4^2 - 0.3542\varphi_1\varphi_2 - 0.7975\varphi_1\varphi_3 - 1.5502\varphi_1\varphi_4 + 2.7841\varphi_2\varphi_3 \\
&- 0.5386\varphi_2\varphi_4 + 0.2573\varphi_3\varphi_4 - 0.1684\varphi_1^3 - 0.5071\varphi_2^3 + 1.4815\varphi_3^3 - 0.2310\varphi_4^3 \\
&+ 0.8761\varphi_1^2\varphi_2 + 1.3092\varphi_1^2\varphi_3 + 0.1184\varphi_1^2\varphi_4 - 0.0996\varphi_1\varphi_2^2 - 0.3868\varphi_1\varphi_3^2 + 0.0888\varphi_1\varphi_4^2 \\
&+ 1.1500\varphi_2^2\varphi_3 - 0.6931\varphi_2^2\varphi_4 - 2.5784\varphi_2\varphi_3^2 + 0.1512\varphi_2\varphi_4^2 - 2.7151\varphi_3^2\varphi_4 + 1.6950\varphi_3\varphi_4^2 \\
&- 2.1074\varphi_1\varphi_2\varphi_3 + 1.1258\varphi_1\varphi_2\varphi_4 + 1.3134\varphi_1\varphi_3\varphi_4 + 0.6595\varphi_2\varphi_3\varphi_4
\end{aligned}
\tag{13}
$$

Similarly, the damping regression equation can be acquired based on the stiffness calculation method:

$$\begin{aligned}\frac{\hat{f}_2}{E11} &= -0.2648 + 3.0819\varphi_1 - 1.6038\varphi_2 - 2.5479\varphi_3 + 2.5604\varphi_4 - 2.1802\varphi_1^2 + 1.4020\varphi_2^2 \\ &+1.5283\varphi_3^2 - 1.6530\varphi_4^2 - 0.8512\varphi_1\varphi_2 - 1.5508\varphi_1\varphi_3 - 3.7226\varphi_1\varphi_4 + 5.4696\varphi_2\varphi_3 \\ &-0.8495\varphi_2\varphi_4 + 0.0787\varphi_3\varphi_4 - 0.1734\varphi_1^3 - 0.9562\varphi_2^3 + 1.4337\varphi_3^3 - 0.2797\varphi_4^3 \\ &+1.6602\varphi_1^2\varphi_2 + 2.1616\varphi_1^2\varphi_3 + 0.8800\varphi_1^2\varphi_4 - 0.1481\varphi_1\varphi_2^2 - 0.3982\varphi_1\varphi_3^2 + 0.6600\varphi_1\varphi_4^2 \\ &+0.9527\varphi_2^2\varphi_3 - 0.8391\varphi_2^2\varphi_4 - 3.9493\varphi_2\varphi_3^2 + 0.4225\varphi_2\varphi_4^2 - 3.5250\varphi_3^2\varphi_4 + 2.5354\varphi_3\varphi_4^2 \\ &-3.3547\varphi_1\varphi_2\varphi_3 + 1.6821\varphi_1\varphi_2\varphi_4 + 1.9625\varphi_1\varphi_3\varphi_4 + 0.3758\varphi_2\varphi_3\varphi_4\end{aligned} \tag{14}$$

Then the stiffness and damping of the dual-chamber air spring can be calculated by the following formulas:

$$K_{\text{modle}} = f_1(\varphi_1, \varphi_2, \varphi_3, \varphi_4) = \hat{f}_1 \cdot E13 \tag{15}$$

$$C_{\text{modle}} = f_2(\varphi_1, \varphi_2, \varphi_3, \varphi_4) = \hat{f}_2 \cdot E11 \tag{16}$$

Equations (15) and (16) determine the cubic polynomial equation of stiffness and damping.

5.3. Test of Calculation Model

In statistics, regression analysis is performed using the variables. The coefficient of determination. R^2 is the ratio between the regression sum of squares and sum of squares for total. R^2 is between 0 and 1. The closer it is to 1, the better the fit of the regression. A goodness of fit of more than 0.8 is typically considered acceptable.

In multiple regression analysis, R^2 expression:

$$R^2 = 1 - \frac{SSE}{SST} \tag{17}$$

where *SST* is the sum of squares for total; and *SSE* is the sum of squares for error.

$$SST = \sum_{i=1}^{N} \left(f_{1i} - \bar{f}_{1i}\right)^2 \tag{18}$$

$$SSE = \sum_{i=1}^{N} \left(f_{1i} - \hat{f}_{1i}\right)^2 \tag{19}$$

The coefficient of determination R^2 is related to the number of independent variables. In order to accurately test the accuracy of the model, it is necessary to take into account the degrees of freedom and make adjustments of R^2 expression according to the size of the formula.

$$R_a^2 = 1 - \frac{SSE/(n-p-1)}{SST/(n-p)} = 1 - \left(1 - R^2\right) \cdot \left(\frac{n-1}{n-p-1}\right) \tag{20}$$

R_a^2 is denoted as the adjusted coefficient of determination; n is the total number of experiments, and p is the number of variables. The adjusted coefficient of determination more accurately reflects the degree of fit between the model and experimental data.

By making an analysis of the data ($n = 49$, $p = 35$), the following can be obtained:

1 For the stiffness model $R^2 = 0.99154$, $R_a^2 = 0.96378$, the goodness of fit is 96.378%, indicating that the model can well simulate and predict the experimental results. This indicates the effectiveness of the cubic polynomial regression equation fit.

2 For the damping model, $R^2 = 0.98970$, $R_a^2 = 0.95470$, the goodness of fit is 95.470%, again indicating that the model can well simulate and predict the experimental results. This again supports the effectiveness of the cubic polynomial regression equation fit.

5.4. Prediction of Experimental Results by the Rapid Calculation Model

To further validate the forecast accuracy of the stiffness and damping rapid calculation model, we conducted a series of experiments. To ensure the universality, these experiments consist of random combinations of four factors at different levels, different from the levels tested in the test programs in the orthogonal table.

The straight lines in Figure 18 and Figure 19 are experimental values, while red dots are predictive values for the rapid calculation model. If a dot falls on the line, this indicates the predicted value is equivalent to the experimental value; deviations from the straight line indicate differences between the predicted and the experimental values. As seen in the figures, the predicted and experimental values fit well. For more horizontal level combination schemes, the predicted value can be calculated based on the rapid calculation model proposed in this paper, which can play a guiding role to engineering structural design and applications.

The experiments performed here were representative in the parameter range, and the polynomial prediction model is appropriate for these values. The application of the polynomial prediction model is as follows: upper and lower air chamber volume ratio, $\xi 1 \in [0.5-2.0]$, orifice ratio, $\xi 2 \in [0-0.6]$, loading amplitude ratio, $\xi 3 \in [0.01-0.07]$, and loading frequency $f \in [0.497-5.474]$.

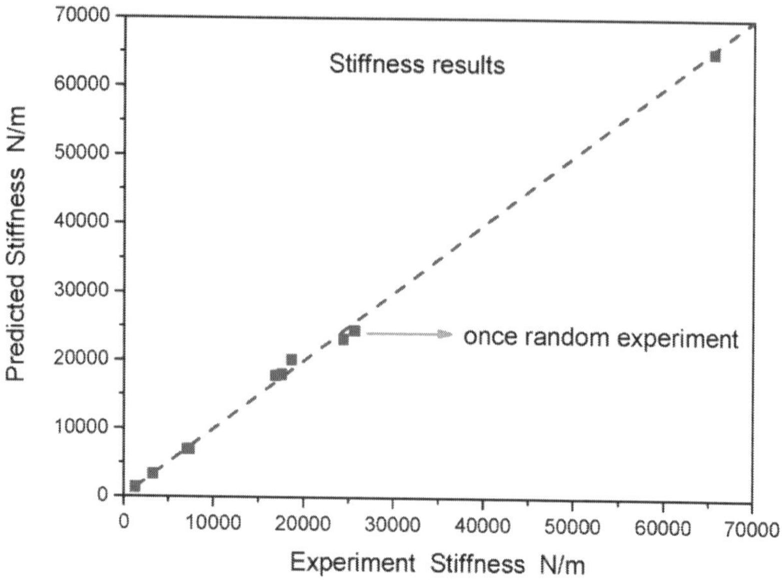

Figure 18. Comparison of experimental and predicted values on the dual-chamber air spring stiffness.

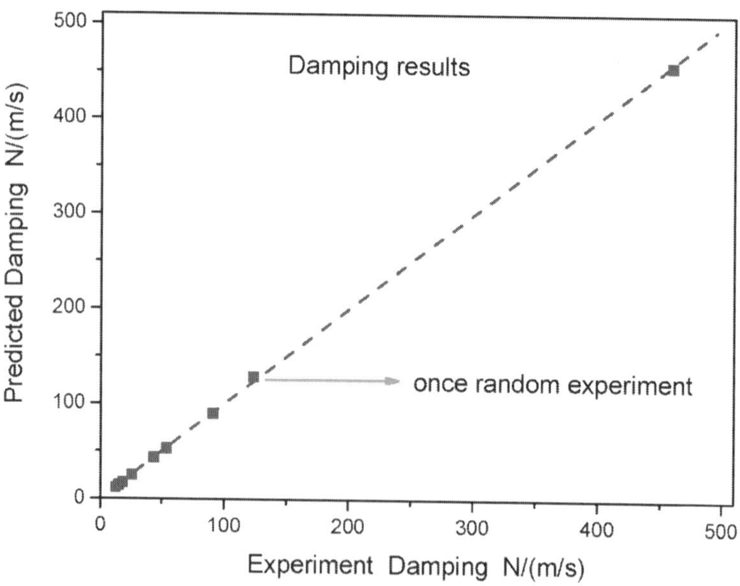

Figure 19. Comparison of experimental and predicted values on the dual-chamber air spring damping.

6. CONCLUSIONS

The mechanical differences between the gas-liquid coupling dual-chamber air spring device and the traditional one make the analysis of effects on stiffness and damping more challenging. The compression of the gas in the traditional air spring is directly on the external load of the air chamber by the piston. The pressure variation of the air spring chamber is the same phase and amplitude as the external load, but for the dual-chamber air spring device applied to motion suppression of marine structures, the movement of the liquid column and the external excitation differ in phases and amplitude.

The current study analyzes the characteristics of a dual-chamber air spring device. A parallel model and series model were used to simulate the dual-chamber air spring energy-absorbing device. An orthogonal test scheme was used to investigate the effect laws of four factors (upper and lower air chamber volume ratio, orifice ratio, loading amplitude, and loading frequency of outer excitation) on the air vibration characteristics of the dual-chamber. Based on the experimental results, a higher-order non-linear regression method was obtained, achieving a rapid calculation model for dual-chamber air spring stiffness and damping, and the reliability of the method was verified experimentally. Using the rapid calculation model, from the upper and lower air chamber volume, orifice ratio, the frequency, and amplitude of external load, we can determine the stiffness and damping of the dual-chamber air spring device applied to motion suppression of marine structures by the formulas (15) and (16). The main findings are:

1 Based on energy consumption results, the goodness of fit of the parallel model was 89.43%, and the goodness of fit of the series model was 99.88%. The parallel model is more consistent with the real physical model.

2 The effects of volume ratio and orifice ratio on dual-chamber vibration absorber stiffness were not monotonic, but the loading amplitude ratio and frequency tended toward monotonic increasing.

3 The effects of the volume ratio, orifice ratio, and loading amplitude ratio on the dual-chamber vibration absorber stiffness did not behave in a monotonic manner, but the loading frequency on damping tended toward monotonic increasing.

4 A polynomial rapid calculation model for stiffness and damping was constructed. The accuracy of the rapid calculation model results was verified by the experimental results, and the predicted values were in good agreement with the experimental values.

ACKNOWLEDGMENTS

The authors would like to express their gratitude for the financial support of the National Natural Science Foundation of China (Grant Nos. 11072246, 51490673, 10702073, 11232012) and the National Basic Research Program of China (973 Program 2014CB046801)

AUTHOR CONTRIBUTIONS

Xiaohui Zeng put forward the overall idea. Liang Zhang contributed to the theoretical and experimental analysis, and wrote the manuscript. Yang Yu contributed to the experimental study. Min Shi contributed to data analysis. Jifu Zhou contributed to data analysis.

CONFLICTS OF INTEREST

The authors declare no conflict of interest.

REFERENCES

1. Taylor, R.E.; Jefferys, E.R. Variability of hydrodynamic load predictions for a tension leg platform. *Ocean Eng.* **1986**, *13*, 449–490.
2. Zeng, X.H.; Shen, X.P.; Wu, Y.X. Governing equations and numerical solutions of tension leg platform with finite amplitude motion. *Appl. Math. Mech.* **2007**, *28*, 37–49.

3. Zeng, X.H.; Li, X.W.; Liu, Y.; Wu, Y.X. Nonlinear dynamic responses of tension leg platform with slack-taut tether.*China Ocean Eng.* **2009**, *23*, 37–48.

4. Ahmad, S.K.; Ahmad, S. Active control of non-linearly coupled TLP response under wind and wave environments.*Comput. Struct.* **1999**, *72*, 735–747.

5. Alves, R.M.; Batista, R.C. *Active/Passive Control of Heave Motion for TLP Type Offshore Platform*; Chung, J.S., Matsui, T., Koterayama, W., Eds.; International Society of Offshore and Polar Engineers: Cupertino, CA, USA, 1999; pp. 332–338.

6. Sakai, F.; Takaeda, S.; Tamaki, T. Tuned liquid column damper-new type device for suppression of building vibrations. In Proceedings of the International Conference on Highrise Buildings, Nanjing, China, 25–27 March 1989; pp. 926–931.

7. Lee, H.H.; Wong, S.H.; Lee, R.S. Response mitigation on the offshore floating platform system with tuned liquid column damper. *Ocean Eng.* **2006**, *33*, 1118–1142.

8. Lee, H.H.; Juang, H.H. Experimental study on the vibration mitigation of offshore tension leg platform system with UWTLCD. *Smart Struct. Syst.* **2012**, *9*, 71–104.

9. Taflanidis, A.A.; Angelides, D.C.; Scruggs, J.T. Simulation-based robust design of mass dampers for response mitigation of tension leg platforms. *Eng. Struct.* **2009**, *31*, 847–857.

10. Lee, S.K.; Lee, H.R.; Min, K.W. Experimental verification on nonlinear dynamic characteristic of a tuned liquid column damper subjected to various excitation amplitudes. *Struct. Des. Tall Spec. Build.* **2012**, *21*, 374–388.

11. Chatterjee, T.; Chakraborty, S. Vibration mitigation of structures subjected to random wave forces by liquid column dampers. *Ocean Eng.* **2014**, *87*, 151–161.

12. Zeng, X.; Yu, Y.; Zhang, L.; Liu, Q.; Wu, H. A New Energy-Absorbing Device for Motion Suppression in Deep-Sea Floating Platforms. *Energies* **2014**, *8*, 111–132.

13. Rijken, O.; Spillane, M.; Leverette, S.J. Vibration Absorber Technology and Conceptual Design of Vibration Absorber for TLP in Ultradeep Water. In Proceedings of the ASME 29th International Conference on Ocean, Offshore and Arctic Engineering, Shanghai, China, 6–11 June 2010; ASME: New York, NY, USA, 2010; pp. 629–638.

14. Bian, X.S.; Leverette, S.J.; Rijken, O.R. A TLP solution for 8000 Ft water depth. In Proceedings of the ASME 29th International Conference on Ocean, Offshore and Arctic Engineering, Shanghai, China, 6–11 June 2010; ASME: New York, NY, USA, 2010; pp. 255–262.

15. Spillane, M.W.; Rijken, O.R.; Leverette, S.J. Vibration absorbers for deepwater TLP's. In Proceedings of the 17th International Offshore and Polar Engineering Conference, Lisbon, Portugal, 1–6 July 2007.

16. Bachrach, B.I.; Rivin, E. Analysis of a damped pneumatic spring. *J. Sound Vib.* **1983**, *86*, 191–197.

17. Lee, J.H.; Kim, K.J. Modeling of nonlinear complex stiffness of dual-chamber pneumatic spring for precision vibration isolations. *J. Sound Vib.* **2007**, *301*, 909–926.

18. Jing, X.; Lang, Z. *Frequency Domain Analysis and Design of Nonlinear Systems based on Volterra Series Expansion: A Parametric Characteristic Approach*; Springer: Berlin, Germany; 2015.

19. Xiao, Z.; Jing, X. An SIMO Nonlinear System Approach to Analysis and Design of Vehicle Suspensions, IEEE/ASME trans. *Mechatronics* **2015**, *20*, 3098–3111.

20. Jing, X. Nonlinear characteristic output spectrum for nonlinear analysis and design. *IEEE/ASME Trans. Mechatron.* **2014**, *19*, 171–183.

21. Shin, Y.H.; Kim, K.J. Performance enhancement of pneumatic vibration isolation tables in low frequency range by time delay control. *J. Sound Vib.* **2009**, *321*, 537–553.

22. Moon, J.H.; Lee, B.G. Modeling and sensitivity analysis of a pneumatic vibration isolation system with two air chambers. *Mech. Mach. Theory* **2010**, *45*, 1828–1850.

23. Li, Z.; Li, X.; Chen, X. Generic vibration criteria-based dual-chamber pneumatic spring vibration isolation table design. *Proc. Inst. Mech. Eng. B J. Eng. Manuf.* **2014**, *228*, 1621–1629.

24. Liu, H.; Lee, J.C. Model development and experimental research on an air spring with auxiliary reservoir. *Int. J. Automot. Technol.* **2011**, *12*, 839–847.

25. Zhu, S.; Wang, J.; Zhang, Y. Research on theoretical calculation model for dynamic stiffness of air spring with auxiliary chamber. In Proceedings of the IEEE Vehicle Power and Propulsion Conference, Harbin, China, 3–5 September 2008.

26. Pu, H.; Luo, X.; Chen, X. Modeling and analysis of dual-chamber pneumatic spring with adjustable damping for precision vibration isolation. *J. Sound Vib.* **2011**, *330*, 3578–3590.

27. Erin, C.; Wilson, B.; Zapfe, J. An improved model of a pneumatic vibration isolator: Theory and experiment. *J. Sound Vib.* **1998**, *218*, 81–101.

28. Heertjes, M.; van de Wouw, N. Nonlinear dynamics and control of a pneumatic vibration isolator. *J. Vib. Acoust.* **2006**,*128*, 439–448.

29. Lee, J.H.; Kim, K.J. A method of transmissibility design for dual-chamber pneumatic vibration isolator. *J. Sound Vib.***2009**, *323*, 67–92.

30. Toyofuku, K.; Yamada, C.; Kagawa, T.; Fujita, T. Study on dynamic characteristic analysis of air spring with auxiliary chamber. *JSAE Rev.* **1999**, *20*, 349–355.

31. Rao, C.R. Factorial experiments derivable from combinatorial arrangements of arrays. *Suppl. J. R. Stat. Soc.* **1947**, *9*, 128–139.

32. Taguchi, G. Performance analysis design. *Int. J. Prod. Res.* **1978**, *16*, 521–530.

33. Azouzi, R.; Guillot, M. On-line prediction of surface finish and dimensional deviation in turning using neural network based sensor fusion. *Int. J. Mach. Tools Manuf.* **1997**, *37*, 1201–1217.

34. Green, P.E.; Krieger, A.M.; Wind, Y. Thirty years of conjoint analysis: Reflections and prospects. *Interfaces* **2001**, *31*, S56–S73.

35. Kutner, M.H.; Nachtsheim, C.; Neter, J. *Applied Linear Regression Models*; McGraw-Hill/Irwin: New York, NY, USA, 2004.

CHAPTER 7

A semi-empirical ship operational performance prediction model for voyage optimization towards energy efficient shipping

Ruihua Lu [n], Osman Turan, Evangelos Boulougouris, Charlotte Banks, Atilla Incecik

Department of Naval Architecture and Marine Engineering, University of Strathclyde, 100 Montrose Street, Glasgow G4 0LZ, UK

ABSTRACT

Voyage optimization is a practice to select the optimum route for the ship operators to increase energy efficiency and reduce Green House Gas emission in the shipping industry. An accurate prediction of ship operational performance is the prerequisite to achieve these targets. In this paper, a modified Kwon's method was developed to predict the added resistance caused by wave and wind for a specific ship type, and an easy-to-use semi-empirical ship operational performance prediction model is proposed. It can accurately predict the ship's operational performance for a specific commercial ship under different drafts, at varying speeds and in varying encounter angles, and then enables the user to investigate the relation between fuel consumption and the various sea states and directions that the ship may encounter during her voyage. Based on the results from the operational performance prediction model and real time climatological information, different options for the ship's navigation course can be evaluated according to a number of objectives, including: maximizing safety and minimizing fuel consumption and voyage time. By incorporating this into a decision support tool, the ship's crew are able to make an informed decision about what is the best course to navigate.

In this study the Energy Efficiency of Operation (EEO) is defined as an indicator to illustrate the ratio of main engine fuel consumption per unit of transport work. Two case studies are carried out to perform the prediction of ship operational performance for Suezmax and Aframax Oil Tankers, and the results indicate that the semi-empirical ship operational performance prediction model provides extremely quick calculation with very reasonable accuracy, particularly considering the uncertainties related to the parameters of interest for the case study data. Within the case studies, the additional fuel consumption caused by the combined hull and propeller fouling and engine degradation is included in the model as a time-dependent correction factor. The factor may assist the ship owner/operator to determine the hull coating selection, and/or the dry-docking and main engine maintenance strategy.

KEYWORDS

Energy efficient shipping; Voyage optimization; Ship operational performance; Fuel saving; Co_2 emission reduction

1. INTRODUCTION

Energy efficient shipping is a prerequisite for the reduction of the Green House Gas (GHG) emissions to the levels anticipated within the next decades. The continuous growth of the world population and the increase number of developing countries led to the increasing dependence of the world economy on the international trade. For 2007, it was estimated that shipping emits 1046 million tonnes of CO_2 from exhaust emissions, accounting for 3.3% of the global CO_2 emission during that year. CO_2 emission from International shipping alone were estimated to account 2.7% of the global CO_2 emission in 2007, and the carbon dioxide emissions from international shipping was projected to triple by the year 2050 (IMO, 2009). These findings alerted the International Maritime Organization (IMO) and led to the introduction of the first maritime energy efficiency regulations that entered into force on the 1st of January 2013 (IMO, 2011). The aim of the regulations is to reduce carbon emissions by decreasing the amount of fuel consumed. This can be achieved by optimizing the ship's design, deploying new energy efficient technologies, or by improving the ship's operation. The regulations require both new and existing ship above 400 GT to have a ship specific Ship Energy Efficiency Management Plan, SEEMP (IMO, 2012).

An additional drive towards a more energy efficient shipping is the requirement to remain competitive within a fierce market. Although marine engines used for commercial shipping use the cheapest type of 'bunker fuel', the cost of IFO 180 has risen sharply with other petroleum products, increasing from $170/t in 2002, and from $230/t in 2005, to nearly $700/t in July 2014 (Bunker Index, 2014). With such high fuel prices, the bunker costs could account for 50–60% of a ship's total operating costs (Wang and Teo, 2013). The rising fuel price has supported the increasing need for energy efficiency to survive in highly competitive and capacity oversupplied shipping market.

It is important to realize that an optimum route cannot only be evaluated in terms of fuel consumption. Normally, the voyage optimization has multiple, often conflicting, objectives, such as: minimizing costs regardless of arrival time; punctual time of arrival; safety; and passenger comfort. In most cases, improving one objective may reduce efficiency of another. Each attribute therefore requires a weighting of importance. For example, some shipping companies 'business models prioritise on-time arrival and shorter transit times over reduced fuel consumption. For other companies, providing a 'green service' has a higher priority.

Most existing techniques and software solutions for voyage optimization extract the ship's operational performance from a database build on results from similar ships (in terms of type and size). However, the performance of each specific ship in various voyage conditions (speed, fouling and propulsion system degradation, and draft) may be quite different, especially under severe weather conditions. This highlights the need for real-time, flexible ship-specific modeling in order to provide increased accuracy of ship operational performance prediction for voyage optimization. Another common disadvantage of many existing voyage optimization software solutions is that they only present to the ship's master the recommended route. The users of the software cannot test their intended route and compare its performance to the software recommended route. As a result, captains may develop mistrust to the recommended route and proceed according to their own judgement.

Voyage optimization software can be evaluated according to:

- Technical status – the accuracy and practicability of ship operational performance prediction.
- User acceptance – the user friendliness.
- Economic performance – the evaluation of fuel saving based on voyage optimization.

These three evaluation principles are also the objectives of the research presented. This paper focuses on the development of an accurate and practical ship operational performance prediction model that can be used to select the optimum routes for minimum fuel consumption, taking into consideration average ship speed, encountering sea states and voyage time.

The ship operational performance model presented in this paper is developed by the modifying Kwon's method (Kwon, 2008) using a case study of ship's operational data (i.e. ship's noon reports) and sea trial data. The Kwon's method (Kwon, 2008) is an empirical method for the prediction of added resistance due to sea state and wave directions. The case study of ship's operational data is taken as the reference for the modified Kwon's method. This modified model can predict the ship's operational performance for a given wave and weather condition at different speeds, drafts and wave encounter angle in a semi-empirical way.

A decision support tool has been developed to select the optimum course according to the users 'preference. The users can influence the selection of the optimized route by providing different weightings to the optimization objectives (see optimum route a–e listed in Fig. 10).

Besides the development of the ship operational performance prediction and the optimum routes selection, a time-dependent fuel consumption increase rate after ship dry-docking has been identified, which may be helpful in monitoring ship fouling and engine degradation condition. The identified fuel consumption rate of increase will further assist shipping companies with planning dry-docking and engine maintenance scheduling.

2. STATE OF THE ART

2.1. Semi-empirical approaches for predicting the added resistance

The prediction of ship total resistance in waves (*RT*) can typically be performed in two steps (ITTC, 2011):

a) Prediction of still water resistance, *RSW*, at speeds of interest.
b) Prediction of added resistance in waves, *RAW*, at the same speeds.

The prediction of ship total resistance in waves is obtained by summing the above mentioned predicted values:

$$R_T = R_{SW} + R_{AW} \tag{1}$$

several methods are available to determine the still water resistance of ships. In the presented analysis the Holtrop and Mennen method (Holtrop and Mennen, 1982) has been used.

The increase in resistance caused by waves, greater than the still water condition, can also be calculated using several methods, including Strip Method, Radiated Energy Method, Rankine Panel Method, Cartesian Grid Method, CFD Method, Experiment Method, Empirical Method, and Semi-empirical Method. In the following section, some of the semi-empirical methods for added resistance prediction are reviewed.

2.1.1. The approximated – Salvesen method

The Salvesen method (Salvesen, 1978) provides a basic formula for the added resistance calculation.

$$R_{AW} = -\frac{i}{2}k \cos \beta \sum_{j=3,5} \xi_j \left\{ F_j^{I*} + \hat{F}_j^{D} \right\} + R_7 \tag{2}$$

where $\hat{F}_j^{I*}$, is the complex conjugate of the Froude-Krilov part of the exciting force and moment, and $\hat{F}_j^{D}$ is very similar to the diffraction part of the existing force F_j^{D}, k is the wave number, β is wave heading direction, and ξ_j is the motion calculated by the strip theory. R_7 is given by

$$R_7 = -\frac{1}{2}\xi_I^2 k \frac{\omega^2}{\omega_e} \cos \beta \int_L e^{-2kds}(b_{33} + b_{22} \sin^2 \beta) dx \tag{3}$$

Where, ξ_I is the incident wave amplitude, b_{33} and b_{22} are the sectional heave and sway damping coefficient, d is the sectional draft and s is the sectional-area coefficient. Details of formula 2 and 3 are presented in Salvesen (1978).

The Salvesen method is able to provide accurate results for the longer waves regions ($L/\lambda<1.5$). Therefore, to extend its use for short wave length regions a correction is added to the original Salvesen method to produce the approximated – Salvesen method (Matulja et al., 2011). The correction contains an approximated formula proposed by Faltinsen et al. (1980):

$$R_{AW} = \frac{1}{2}\rho g \left(1 + \frac{2\omega U}{g}\right) \int_{L1} \sin^2 \nu n_1 dl \tag{4}$$

where, $L1$ is non shadow zone of the water plane area, U is ship speed, ω is Encounter frequency, n_1 is X component of the inward normal n to the water line, and v is the angle between the tangent to the water line and the x axis.

The final step of the approximated – Salvesen method is:

$$R = a \quad \text{for} \quad L/\lambda \leq 1 \tag{5}$$

$$R = a + b \quad \text{for} \quad 1 < L/\lambda \leq 2 \tag{6}$$

$$R = b \quad \text{for} \quad L/\lambda > 2 \tag{7}$$

where,

$$a = -\frac{i}{2} k \cos \beta \sum_{j=3,5} \xi_i \left\{ F_j^{I*} + \hat{F}_j^D \right\} \tag{8}$$

$$b = \frac{1}{2} \rho g \left(1 + \frac{2\omega U}{g}\right) \int_{L1} \sin^2 v n_1 \, dl \tag{9}$$

However, formula 8 is only valid in wave heading direction $\beta \approx 180^o$ if the speed U is high, and for $\beta \approx 90^o$ if the speed U is low. The full range of heading directions experienced by a ship in practice are therefore not accounted for with this method. This is a disadvantage for accurate ship performance prediction. Furthermore, the definition of the term ‘short waves region' is loosely defined as it is related to the ship length, further increasing the uncertainty in the added resistance prediction.

In an overall view of the approximated – Salvesen method, the Faltinsen's approximated formula was used to evaluate wave reflection added resistance and then the Salvesen's results were combined with the wave reflection added resistance in a semi-empirical way.

2.1.2. Fuji-Takahashi method

The Fuji-Takahashi method (Fuji and Takahashi, 1975) is a semi-empirical method considering the drift force acting on an upright barrel and then correcting these forces with a coefficient for ship shape & speed. The drift force is calculated using the following equation:

$$D=\frac{1}{2}\rho g\xi_a^2\int_{-B/2}^{B/2}\sin^2\beta dy \tag{10}$$

where, ξ_a is the amplitude of the incident waves, ρ is sea water density, g is gravitational acceleration, β is an inclination angle for x-axis on the hull, B is ship breadth.

The added resistance generated by the reflected wave R_{AW} is calculated by:

$$R_{AW}=\alpha_1(1+\alpha_2)\frac{1}{2}\rho g\xi_a^2\int_{-B/2}^{B/2}\sin^2\beta dy \tag{11}$$

$$\alpha_1=\frac{\pi^2 I_1(kd)^2}{\pi^2 I_1(kd)^2+K_1(kd)^2} \tag{12}$$

$$\alpha_2 = 5\sqrt{F_n} \tag{13}$$

where, α_1, α_2, d, and k denote coefficients for the draft effect, ship speed effect, draft, wave number, and F_n is Froude number.

The semi-empirical formula of Fuji-Takahashi is a widely used to predict ship added resistance, but the added resistance due to reflected waves acting on the bulbous-bow is not included. This may lead to significant error in predicting the added resistance as bow flare above the water surface may change the reflected wave properties, along with the presence of the bulbous bow under water.

2.1.3. Kuroda-Tsujimoto-Fujiwara-Ohmatsu-Takagi method

Based on the investigation of Fuji and Takahashi's (1975) semi-empirical method, Kuroda et al. (2008) proposed an improved expression for the added resistance due to wave reflection (R_{AW}). The formula is as follows:

$$R_{AW} = \frac{1}{2}\rho g \xi_a^2 \alpha_d (1+\alpha_U) B B_f(\chi) \tag{14}$$

where, α_d is effect of draft and frequency, $1+\alpha_U$ is effect of advance speed, B_f is bluntness coefficient, and χ indicates ship heading direction.

Kuroda et al. modified the terms for added resistance due to wave reflection by taking into account the effect of draft, wave encounter frequency and speed of advance. The oblique waves were also applied on the method. *Kuroda-Tsujimoto-Fujiwara-Ohmatsu-Takagi method* requires tank testing in short waves and therefore the effect of hull form above water line is captured in the added resistance calculation. However, the added resistance due to reflected waves acting on bulbous-bow is still not included.

2.1.4. A simplified method to calculate added resistance based on Gerritsma and Beukelman's method

Gerritsma and Beukelman's method (Gerritsma and Beukelman, 1972) considers radiated energy to calculate added resistance.

The added resistance is calculated with the following expression:

$$R_{AW} = \frac{-k \cos \beta^2}{2\omega_e} \int_0^L b' |V_{Z_b}|^2 \wp x_b \tag{15}$$

where, k is wave number, β is heading angle, ω_e is frequency of encounter, L is ship's water line length, $|V_{Zb}|$ is the amplitude of the velocity of water relative to the strip, b' is the sectional damping coefficient for speed, x_b is x coordinate on the ship.

The Gerritsma and Beukelman's method (Gerritsma and Beukelman, 1972) is one of the most widely used added resistance modeling methods that utilize Strip Theory, and it provides an accurate added resistance prediction across the different ship types. However, for added resistance calculation, calculating ship motions by strip theory is complex and might be unnecessary in some applications. To address this issue a further simplified method developed by Alexandersson (2009) was proposed for the added resistance calculation. The simplified method uses the Gerritsma and Beukelman's method, applying to a large series of case studies to determine the added resistance, then using linear regression, a series of simplified formulas to determine the added resistance are derived. This simplified method is therefore a semi-empirical method as it uses the results from the Gerritsma and Beukelman's analytical method (Gerritsma and Beukelman, 1972) and combines it with regression techniques.

Although the semi-empirical method simplifies the more complicated strip theory calculations whilst still providing relatively stable predictions for added resistance, the limitations of the method still exist, such as the prediction accuracy decreases in bow/beam/following waves and high frequency waves; trim of the ship is not included.

2.1.5. Summary of semi-empirical methods to predict added resistance

In an overview of the existing semi-empirical methods for added resistance prediction, they either combine empirical methods with analytical results, or update existing empirical method with analytical method, in a semi-empirical way. The common disadvantage for these methods is that they are not able to predict the added resistance accurately with different encounter angle. The short or long wave length is also a critical parameter for selecting the right formulas. The semi-empirical methods, which involve tank tests, may sharply increase the modeling cost as it may take much longer time for added resistance estimation.

2.2. Voyage optimization in routing service

Extensive surveys of research on ship routing and scheduling have been carried out about every 10 years (Ronen, 1983, Ronen, 1993, Christiansen et al., 2004 and Christiansen et al., 2013). Fagerholt et al. (2010) proposed mathematical models to optimize speed on a ship route. However, the fuel consumption was approximated by a cubic function, which is not accurate enough for voyage optimization.Padhy et al. (2008) predicted the speed loss due to weather condition using sea-keeping computing tools. The pre-computed Response Amplitude Operator (RAO) was employed in Dijkstra's algorithm to obtain optimum route in a given sea-state. Hinnenthal and Glauss (2010) utilized strip-theory and wave spectra to generate RAO, and predicted the added resistance through a statistical evaluation method for voyage optimization.Avgouleas (2008) estimated mean added resistance in waves with the aid of SWAN1, which is an advanced frequency domain CFD code using Rankine Panel Methods. Then the Iterative Dynamic Programming algorithm was employed to achieve voyage optimization. The ship's response based routing and high fidelity computational hydrodynamic performance based routing require huge amount of computations, long simulation time and the up-to-date ship conditions are not involved. It seems to be immature for accurate and efficient voyage optimization. Larsson and Simonsen (2014), and Shao (2013) adopted Kwon's method (Kwon, 2008) to predict added resistance for weather routing. However, Kwon's method is a generic approach for large number of commercial ship types. Therefore, regarding the limited accuracy of added resistance and ship operational performance modeling for specific commercial ship, the accuracy of voyage optimization can be further increased.

The ship routing service can generally be categorized into ashore based routing services, on-board based routing services and the combination of ashore and on-board routing services. Table 1 provides an overview of available ship routing service.

Table 1. Exemplary compilation of routing service or decision support systems (Hinnenthal and Clauss, 2010).

Service provider	Installed Location	Service/System	Weather forecast	Route planning	Route optimization	Ship monitoring	Data recording
Aerospace and Marine International (USA)	ashore	Weather 3000, internet service, maps displaying fleet and weather information	X				X
Weather Routing Inc. (USA)	ashore	routing advice and Dolphin navigation program combined with a web-based interactive site	X	X			
Finish Meteorological Institute (Finland)	ashore	weather and routing advice for the Baltic sea	X	X			
Fleetweather (USA)	ashore	Meteorological consultancy	X	X			X
Metworks Ltd. (UK)	ashore	meteorological consultancy	X	X			
Applied Weather Technology (USA)	on-board	BonVoyage System	X	X			
Euronav (UK)	on-board	seaPro, software or fully integrated bridge system	X	X			X
Germanischer Lloyd, Amarcon B.V. (Germany, Netherlands)	on-board	SRAS – Shipboard Routing Assistance System	X	X		X	
Transas (UK)	on-board	ship guard SSAS, software or integrated to bridge system	X	X		X	X
Norwegian met office, C-Map (Norway, Italy)	on-board	C-STAR	X	X			
US Navy (USA)	on-board	STARS	X	X	X	X	
Meteo Consult (Netherlands)	on-board	SPOS - Ship Performance Optimization System	X	X	X	X	
Oceanweather INC., Ocean Systems INC. (USA)	on-board	VOSS – Vessel Optimization and Safety System	X	X	X	X	
Weather News International, Oceanwaves (USA, Japan)	ashore & on-board	voyage planning system VPS and ORION, routing and optimization software	X	X	X		
Swedish Met and Hydrology Institute (Sweden)	ashore & on-board	Seaware Routing, Seaware Rouring Plus and Seaware EnRoute Live	X	X		X	
Deutscher Wetterdienst (Germany)	ashore & on-board	MetMaster, MetFerry, routing system, advice on demand	X	X	X		

3. DATA DESCRIPTION

Ships have to collect operational data on a daily basis, known as ship logs and ship reports (often referred to as noon reports as they are typically recorded every 24 h at noon). The type of data fields that are included in the ship reports cover: date/time of the report, ship position, and estimated time of arrival, arrival/departure port, observed distance, achieved speed, mean draft, Beaufort Number, wind direction, and total main engine fuel consumption per day. There is no standard for the recording of operational parameters within the ship reports and therefore the content tends to differ between companies compared to the parameters mentioned above. These parameters contain a vast amount of uncertainty. This uncertainty originates from the methods used to obtain their measurement, the type of measurement, human error and the assumptions made during analysis. Some of the uncertainties related to the parameters of interest for the case study data are discussed here:

- Achieved speed: the achieved speed is calculated by dividing the observed distance recorded by the report duration. The speed is therefore given as an average value for the whole report duration. It does not take into account the speed profile which, due to the approximately cubic relationship between ship speed and power, could have a significant influence over the fuel consumed during the reporting period. The achieved speed is also the speed over ground and therefore, the effects of currents and tides are not taken into account. To improve the accuracy of performance prediction, the speed through water should be obtained.
- Beaufort Number (BN): the Beaufort measurement itself contains uncertainty as one number is used to represent a range of wave heights and sea conditions. More accurate added resistance performance prediction methods depend on the wave height as an input along with the type of sea spectrum (including surface waves and developed seas). Additional uncertainty is created with the measurement of Beaufort Number being made via judgement of the sea conditions typically out of the window on the bridge by the officer on watch. Not only is this measurement subjective as it is a judgement, it is also observed from some distance away from the sea surface. There is also ambiguity as to whether the Beaufort Number recorded is representative of the conditions at the observation point, or an average of the conditions observed over the report duration.
- Wind direction: recording of the wind direction is typically aided by the use of an anemometer. Obstructing super structure in different wind directions is known to produce inaccuracies in the measurement, along with variations in wind strength at different heights. Uncertainty due to averaged measurements also applies in the same way as for Beaufort Number. Furthermore, the wind direction is assumed to be the same as the wave direction, which may be true in most instances of surface waves, but it could also be very different for swell direction. To improve sea and wind condition measurements the wave, swell and wind direction and strength or height, should be recorded.
- Ship heading direction: the angle between the direction of ship bow and the North Pole. The angle is measured clockwise from north, in degrees from 0° to 359°.

- Encounter angle: derived from wind direction, which is relative to the ship. It is also known as the weather direction, as presented in Fig. 1.

- Main engine fuel consumption: fuel flow meters improve the accuracy of fuel consumption measurements if they are calibrated and working correctly. However, in most cases the main engine fuel oil consumption is recorded by tank sounding. Not only could the measurement contains a vast amount of inaccuracy, but there is room for error in the tank sounding calculations and the recorded value is susceptible to transcription error and intentional alteration for various reasons.

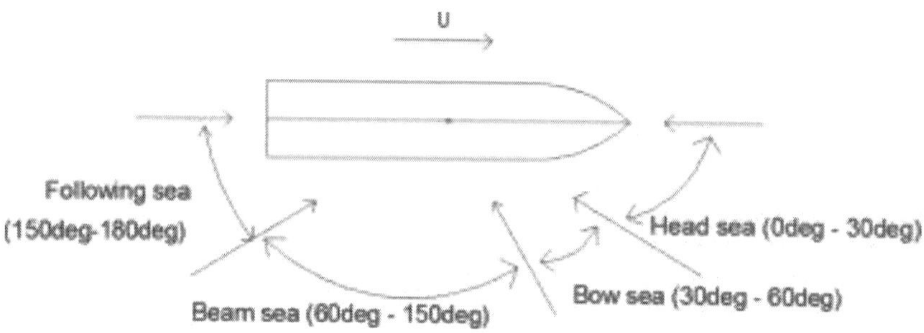

Figure 1. Encounter angle.

Despite all of the uncertainties described, the parameters in the ship reports provide an insight into the operating conditions of the ship in sailing and thus provide a value to performance prediction modeling.

4. MODEL DESCRIPTION

The method for a new semi-empirical approach proposed within this paper that can be used for modeling ship operational performance is introduced in this section. A modified Kwon's method is developed to enhance the accuracy of added resistance prediction. This method is based on Kwon (2008) added resistance modeling method but is updated to take into account ship specific characteristics by utilizing the analysis of collected operational data. The first step of the semi-empirical method is the estimation of the still water resistance. This is followed by the prediction of added resistance due to wind and wave conditions.

4.1. Still water resistance modeling

The well-known Holtrop and Mennen's Method (Holtrop and Mennen, 1982) is used to estimate the still water resistance of the ship. This method is widely used to calculate the total still water resistance of a ship with a good accuracy for a wide range of ship types, sizes, hull forms and for a range of Froude numbers.

4.2. Added resistance modeling

Kwon (2008) added resistance model is an approximate method for predicting speed loss of a displacement type ship due to added resistance in weather conditions (irregular waves and wind). The advantage of this method is that it is easy and practical to use.

The weather effect, presented as speed loss, compares the speed of the ship in varying actual sea conditions to the ship's expected speed in still water conditions. It is expressed in the following way using Kwon's method for modeling added resistance (Kwon, 2008):

$$\frac{\Delta V}{V_1} 100\% = C_{\beta} C_U C_{\mathrm{Form}} \tag{17}$$

$$V_2 = V_1 - \left(\frac{\Delta V}{V_1} 100\%\right) \frac{1}{100\%} V_1 = V_1 - (C_{\beta} C_U C_{\mathrm{Form}}) \frac{1}{100\%} V_1 \tag{18}$$

where,

V_1: Design (nominal) operating ship speed in still water conditions (no wind, no waves), given in m/s.

V_2: Actual ship speed in the selected weather (wind and irregular waves) conditions, given in m/s.

$\Delta V = V_1 - V_2$ Absolute speed loss, given in m/s.

C_β: Direction reduction coefficient, dependent on the weather direction angle (with respect to the ship's bow) and the Beaufort Number (BN), as shown in Table 2.

C_U: Speed reduction coefficient, dependent on the ship's block coefficient C_B. The loading condition and the Froude Number (F_n), as shown in Table 3.

C_{form}: Ship form coefficient (C_{form}), as shown in Table 4.

Table 2. Direction reduction coefficient C_β due to weather direction.

Weather direction	Encounter angle (deg)	Direction reduction coefficient C_β
Head sea (irregular wave) and wind	0–30	$2C_\beta = 2$
Bow sea (irregular wave) and wind	30–60	$2C_\beta = 1.7 - 0.03((BN-4)^2)$
Beam sea (irregular wave) and wind	60-150	$2C_\beta = 0.9 - 0.06((BN-6)^2)$
Following sea (irregular wave) and wind	150–180	$2C_\beta = 0.4 - 0.03((BN-8)^2)$

Table 3. Speed reduction coefficient Cu due to Block coefficient C_b.

Block coefficient C_b	Ship loading conditions	Speed reduction coefficient Cu
0.55	normal	$1.7 - 1.4Fn - 7.4F_n{}^2$
0.6	normal	$2.2 - 2.5Fn - 9.7F_n{}^2$
0.65	normal	$2.6 - 3.7Fn - 11.6F_n{}^2$
0.7	normal	$3.1 - 5.3Fn - 12.4F_n{}^2$
0.75	loaded or normal	$2.4 - 10.6Fn - 9.5F_n{}^2$
0.8	loaded or normal	$2.6 - 13.1Fn - 15.1F_n{}^2$
0.85	loaded or normal	$3.1 - 18.7Fn + 28.0F_n{}^2$
0.75	ballast	$2.6 - 12.5Fn - 13.5F_n{}^2$
0.8	ballast	$3.0 - 16.3Fn - 21.6F_n{}^2$
0.85	ballast	$3.4 - 20.9Fn + 31.8F_n{}^2$

Table 4. Ship form coefficient C_{form} due to ship categories and loading condition.

Type of (displacement) ship	Ship form coefficient
All ships(except container ships) in loaded loading condition	$0.5BN+BN^{6.5}/(2.7\times\Delta^{2/3})$
All ships(except container ships) in ballast loading condition	$0.7BN+BN^{6.5}/(2.7\times\Delta^{2/3})$
Container ships in normal loading conditions	$0.7BN+BN^{6.5}/(22\times\Delta^{2/3})$

The Kwon's method (Kwon, 2008) provides a general introduction to the calculation of ship speed loss in different weather and sea conditions based on the ship's hull form, encounter angle and sea state. However, the method is not able to provide a very accurate prediction of added resistance for each specific ship. Therefore, the modified Kwon's added resistance modeling method developed and presented in this paper includes unique direction reduction coefficients, speed reduction coefficients and ship form coefficients for specific ship type and size. These coefficients are determined from the analysis of the recorded ship operational data. The case studies in Section 6 will be used to verify that the modified Kwon's method is a practical way to provide increased accuracy for the prediction of a specific ship's fuel consumption at varying speeds, encounter angles, drafts and sea states.

4.3. Ship operational performance modeling

Since the ship still water resistance has been predicted utilizing Holtrop and Mennen's Method (Holtrop and Mennen, 1982), the relation between vessel speed (U), total still water resistance (R_{total}), and the required effective power (P_E) can be extracted.

$$P_E = R_{total} * U \tag{19}$$

As the added resistance has been modeled by utilizing the modified Kwon's method, the speed loss under varying drafts, Beaufort Number (BN) and encounter angles has been modeled for specific commercial ship. The corresponding original still water speed (V_{cosw}) can be calculated by combining the actual ship speed (V_{actual}) in a seaway and speed loss (V_{loss}) due to added resistance.

$$V_{cosw} = V_{actual} + V_{loss} \tag{20}$$

Thus, under specific BN and ship heading direction, there is a corresponding original still water speed (V_{cosw}) for an actual ship speed (V_{actual}). Based on formula 19, for each still water speed, there is a corresponding required effective power. The relationship between actual ship speed (V_{actual}) and required effective power (P_E) can be extracted.

From effective power the required brake power (P_B) of main engine can then be determined as:

$$P_B = \frac{P_E}{\eta_T} \tag{21}$$

where η_T is the total power transmission efficiency (from brake power of main engine to effective power):

$$\eta_T = \eta_H * \eta_O * \eta_R * \eta_S \tag{22}$$

where, η_H is the hull efficiency; η_O is the open water efficiency; η_R is the relative rotative efficiency; η_S is the shaft efficiency.

The four efficiencies above can be generally estimated through empirical formulae. However, from the point of view of accurate operational performance prediction for specific commercial ship, the utilization of Speed-Power Curve in the sea trial documents can be used to extract total power transmission efficiency. When sea trial documents are available, for each specified speed, the total power transmission efficiency can be calculated using the Holtrop and Mennen's method (Holtrop and Mennen, 1982) to determine the effective power and the Speed-Power Curve from sea trial documents to read the corresponding brake power of main engine. Since the total power transmission efficiency has been determined, the relationship between actual ship speed and required engine power under varying sea states is generated.

For each specific ship, the corresponding main engine performance documents contain the expected fuel consumption based on ISO reference conditions, which illustrate the Specific Fuel Oil Consumption (SFOC) with corresponding engine load, engine power and engine speed. The ship main engine Fuel Consumption Rate (FCR) can be determined as:

$$FCR = P_B * SFOC \tag{23}$$

Finally, the ship main engine fuel consumption rate under varying speeds, sea states, drafts, and ship heading directions can be predicted.

4.4. Weather and sea state modeling

To use the performance prediction model described as part of a voyage optimization model, a source of weather and sea state forecasting needs to be identified as an input. A 'GRIB2' ocean weather forecast file from the National Oceanic and Atmospheric Administration (NOAA) (2015) is used for this purpose. The decoding program has been written in house to read and output the global ocean weather forecast, as presented inFig. 2. The information contained in the file includes significant wave height, swell, wind speed, and directions.

A flow diagram of the semi-empirical method proposed in this paper to predict the ship operational performance is shown in Fig. 3. The dashed boxes on the left indicate the inputs for this proposed model, and the modeling steps are shown in the solid boxes. The relationship flow between the inputs and modeling steps has been illustrated with arrows.

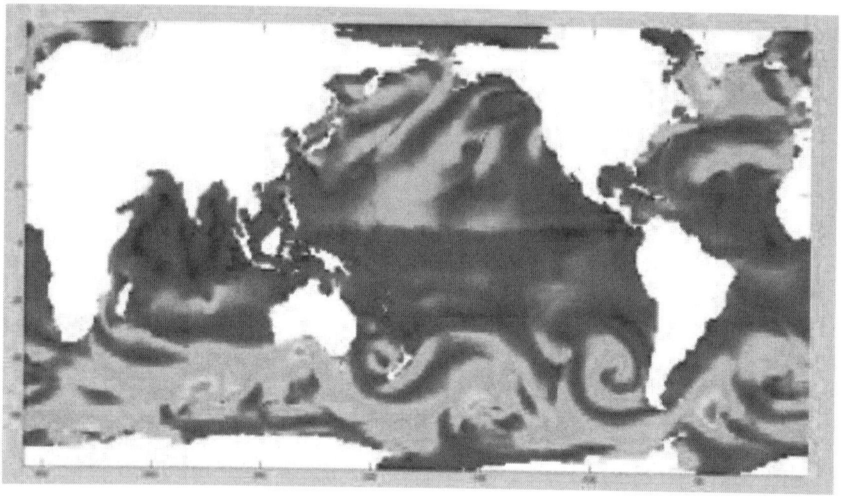

Figure 2. Screenshot from decode program, graph of global ocean weather forecast.

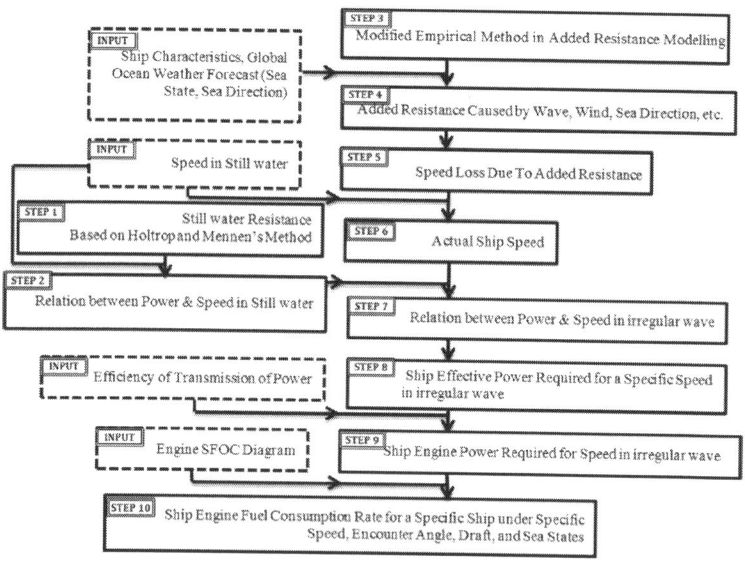

Figure. 3. Analysis diagram of the proposed semi-empirical ship operational performance prediction model.

5. VOYAGE OPTIMIZATION

In this section, the development of grids system and utilization of the proposed performance prediction model in selecting the optimum route are illustrated. At this stage, the optimum route is defined as the one with minimum fuel consumption regarding given average ship speed, encountering sea states and voyage time.

5.1. Setting up grids

A grid system is able to clearly present the potential routes on ship navigation charts. At present, the grid system was plotted on world map, which is developed using the Google maps API (Google maps API, 2015). From start point to destination point, each possible route is equally divided into $n+1$ legs by n stages; the nodes in one stage are equally distributed with unique longitude. On each stage, the distance between the adjacent nodes is Δx, as presented in Fig. 4. The number of the stages and the quantity of nodes in each stage are determined with specific voyage area, total distance of route and the availability of computing capacity. Since the grids system for specific voyage has been set up, and the departure time and ship speeds during each stage are decided by users, the weather information for the corresponding time at which the ship is expected in that area will be downloaded into each node by the in house program. Thus the information at each node includes its location (latitude and longitude) and the weather and sea forecast information. The combination of sea direction and ship heading direction at two consecutive nodes determines the encounter angle.

5.2. Route selection

As mentioned in Section 1, the user's preferences are incorporated into the development of the decision support system by using a process of weighting the attributes that are most important to them (e.g. passage time, fuel consumption). Commonly, the route with minimum fuel consumption is very popular in route selection. Based on the grids system, a ship performance model developed in house is able to predict the total main engine fuel consumption of each potential route. The outputs of this model also include the estimated time of arrival (ETA), sea state, encounter angle, and average speed. At this stage, the development of automatic optimization implements in progress. Depending on the preferences/priorities of shipmasters, such as lowest BN, shortest ETA, encounter angle with most head sea and bow sea, or the combination of these objectives with different weightings, we are able to manually select the route with minimum fuel consumption considering weather and sea conditions. One case study of the optimum route selection has been carried out, and the results are presented in Fig. 10. Based on the grids system presented in Fig. 4, a simulation of the ship performance has been carried out. The simulation was run on a typical desktop PC with a 3.4 GHz Intel i7 CPU in serial model. With 36 nodes and 30625 potential routes, and the simulation time was around 1.5 h.

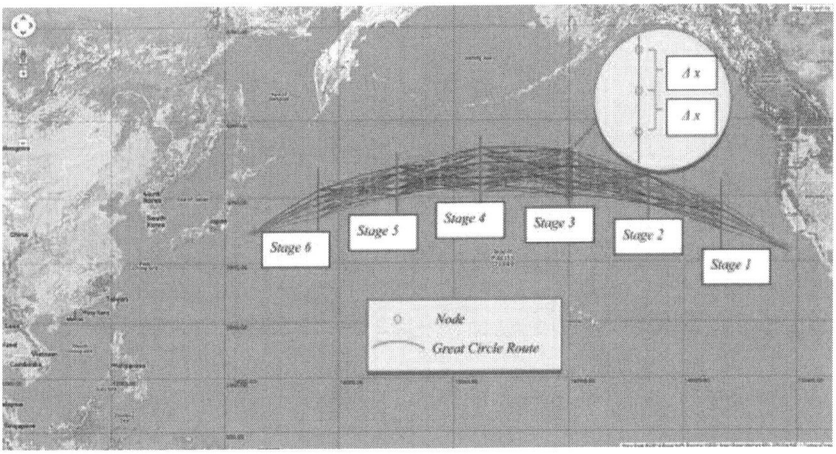

Figure 4. Grids example for the route between Los Angeles Offshore and Chiba, Japan.

6. RESULTS AND DISCUSSION

6.1. Case studies of semi-empirical ship operational performance model

In this study, the 'Energy Efficiency of Operation' (EEO) is defined as the indicator used to illustrate the main engine fuel consumption efficiency and the ship's operational performance.

$$EEO = \frac{FC}{m_{cargo} \times D} \quad (24)$$

where, FC is the main engine fuel consumption (tonnes), m_{cargo} is cargo carried (tonnes), and *D* is the distance in nautical miles corresponding to the cargo carried or work done.

An advantage of using the EEO as an indicator is that it contains many of the same elements and could be easily converted to the Energy Efficiency Operational Index (EEOI), which is recommended within the Ship Energy Efficiency Management Plan (SEEMP) (IMO, 2012).

The basic expression for EEOI for a voyage is defined as:

$$EEOI = \frac{\sum_j FC_j \times C_{F_j}}{m_{cargo} \times D} \quad (25)$$

Where, j is the fuel type, FC_j is the mass of consumed fuel j at one voyage, and C_{Fj} is the fuel mass to CO_2 mass conversion factor for fuel j.

In order to verify the accuracy of the semi-empirical ship operational performance model, the predicted EEO based on the modified Kwon's method and the recorded EEO using recorded operational data from noon reports were compared in the following two case studies. The predicted EEO and the recorded EEO were compared under same conditions, such as Beaufort Number, speed and encounter angle.

6.1.1. Case study 1 – 'Suezmax oil tanker A'

A comparison between the predicted EEO (using the modified Kwon's method) and recorded EEO (using data from the noon reports) for the 'Suezmax oil tanker A' is shown in Fig. 5.

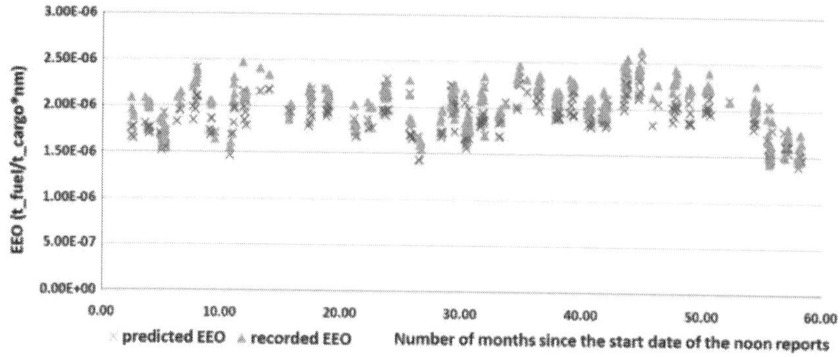

Figure 5. Comparison between predicted EEO and recorded EEO of 'Suezmax oil tanker A'.

The average difference between the predicted EEO and recorded EEO of 'Suezmax oil tanker A' by using the developed method is 5.12% compared to the average difference of 14.7% using the original Kwon's method. Under each sea state (sorted by Beaufort number), the predicted EEO and recorded EEO for each weather direction are compared: a good agreement between predicted and recorded value can be observed as shown in Fig. 6 using BN=3 as an example.

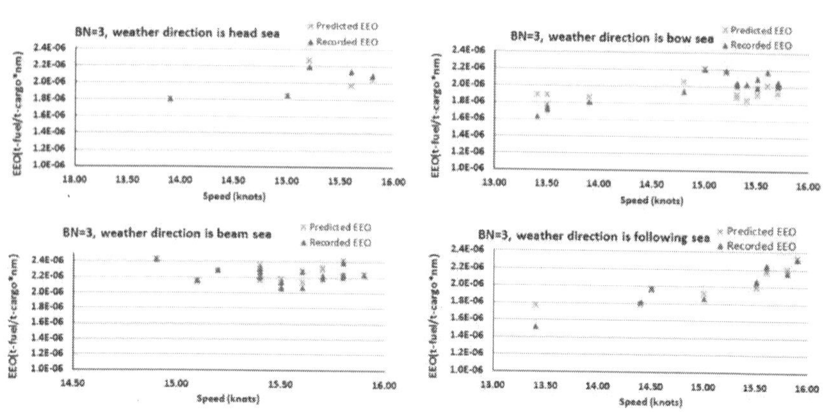

Figure 6. Comparison between predicted EEO and recorded EEO of 'Suezmax oil tanker A' with each weather direction under the BN=3.

6.1.2. Case study 2 – 'Aframax oil tanker B'

The comparison between the predicted EEO and recorded EEO for the 'Aframax oil tanker B' is shown in Fig. 7.

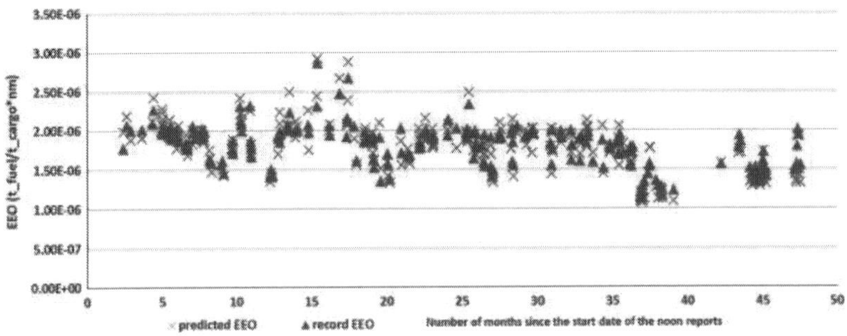

Figure 7. Comparison between predicted EEO and recorded EEO of 'Aframax oil tanker B'.

The average difference between predicted EEO and recorded EEO of 'Aframax oil' tanker by using the developed method is 7.15% compared to the average difference of 21.6% using the original Kwon's method. Under each sea state (sorted by Beaufort number), the predicted EEO and recorded EEO for each wind direction are compared. A good agreement between predicted and recorded value can be observed as shown inFig. 8 using BN=4 as an example.

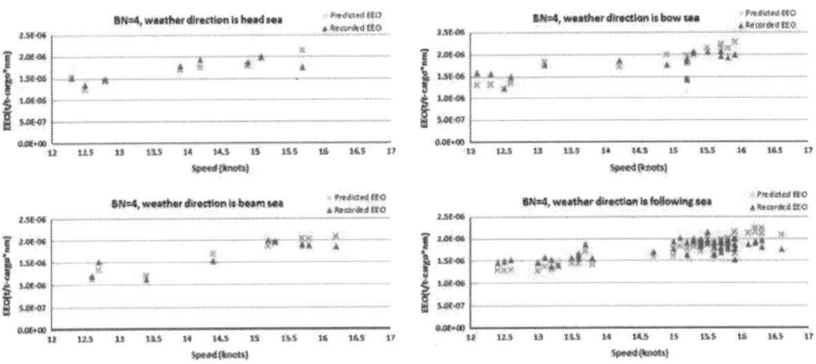

Figure 8. Comparison between predicted EEO and recorded EEO of 'Aframax oil tanker B' with each weather direction under the BN=4.

6.2. Fouling effect and engine degradation

Through the comparison between the predicted EEO and recorded EEO in case study 1 and case study 2, it has been verified that the proposed semi-empirical ship operational performance model provides very reasonable prediction results. However, two factors that are not taken into account within the semi-empirical ship operational performance model are ship hull and propeller fouling effect and main engine degradation conditions. These factors are likely to be a source of error causing the predicted fuel consumption rate using the semi-empirical method to be lower than the rate of fuel consumption recorded in the noon reports. This error in prediction can be taken as the percentage increase in fuel consumption due to hull and propeller fouling and engine degradation, which are known to increase over time between maintenance periods. For the case study of Suezmax oil tanker A and Aframax oil tanker B, the errors are presented in Fig. 9.

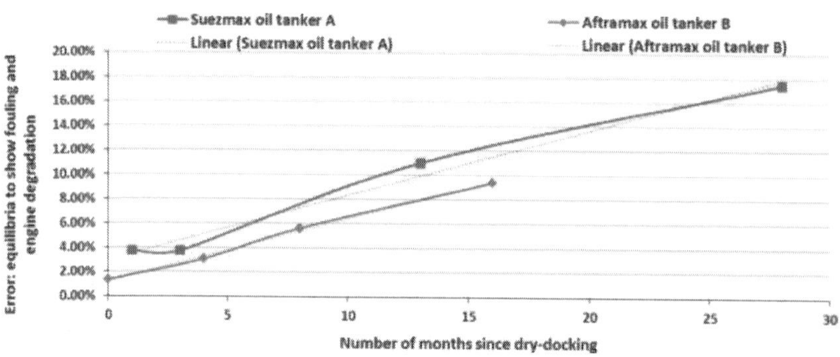

Figure 9. Error between predicted total fuel consumption of each voyage and recorded one since dry-docking for both Suezmax oil tanker A and Aframax oil tanker B.

The trendline A and trendline B indicate that there is a time-dependent factor that increases the total recorded fuel consumption and thus decrease the energy efficiency of the ships. This time-dependent factor can predominantly be assumed to be due to the effect of hull and propeller fouling together with engine degradation. As shown in Fig. 9, the coefficient of determination of trendline A is 0.984, and that of trendline B is 0.9961, both trendline A and B follow very similar trends and increase rates. This illustrates a common rate of the increase in fuel consumption rate caused by hull and propeller fouling and engine degradation. Currently, due to the lack of ship dry-docking reports and main engine maintenance reports, we cannot separate the time-dependent fuel consumption increase rate due to hull and propeller fouling and engine degradation separately. However, the identification of this time-dependent increase in fuel consumption rate can help us to provide up-to-date information about the combined influence on ship performance. The engine data on engine degradation is extremely difficult to obtain and therefore it is a very challenging task to separate the engine degradation and ship fouling and remains a future task to develop a separate modeling.

6.3. Case study of routes selection

With a given departure date and time, draft, fixed average speed, specific ship noon reports and sea trial data, the recommended route with minimum fuel consumption can be identified to the shipmaster as follows (Fig. 10):

- Route a – the blue route is the route with lowest Beaufort Number (low risk to damage the ship and/or its deck cargo; high comfort to passengers) and low fuel consumption.
- Route b – the green route is the Great-Circle Route – with shortest distance between two ports on earth as well as the route with shortest time.
- Route c – the violet route is the route with most head sea and bow sea.
- Route d – the brown route is the route with lowest fuel consumption regardless of voyage time.
- Route e – the red route is the frequently used route as recorded in noon report.

The ship operational performances of these five selected routes have been compared, as shown in Table 5 by using the developed model. The encountered Beaufort Number (BN) and Heading Direction of each route have been listed. The fuel consumption of the selected optimum routes can achieve 10% less than the recorded route. As the average voyage speed is fixed, the voyage durations of the selected optimum routes are very close, but much less than the recorded route.

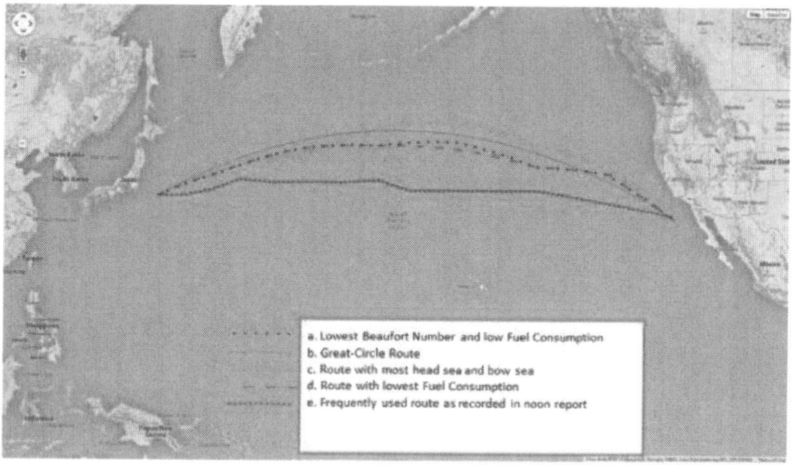

Figure 10. Optimum route selection based on shipmasters 'preference.

Table 5. Comparison of ship operational performance between the selected optimum routes and recorded route.

The encountered Beaufort Number (BN), Heading Direction with given departure date & time, loading condition and fixed average speed	Route a		Route b		Route c		Route d		Route e	
	BN	Direction	BN	Direction	BN	Direction	BN	Direction	BN	Direction
	5	Bow	5	Bow	5	Head	5	Bow	5	Bow
	5	Beam	5	Beam	5	Bow	5	Beam	7	Bow
	3	Bow	4	Bow	5	Bow	3	Bow	6	Head
	3	Beam	3	Beam	4	Beam	3	Beam	5	Head
	3	Beam	4	Beam	4	Beam	3	Beam	5	Head
	3	Beam	3	Beam	4	Bow	3	Beam	5	Head
	1	Head	1	Head	2	Head	1	Head	2	Bow
Voyage Duration (h)	367.7		366.1		368.5		367.3		392	
Main Engine Fuel Consumption (t)	555.5		558.3		580.4		554.9		623.5	
% of Fuel saving compared to Route e	10.90		10.46		6.91		11.01		0	

7. CONCLUSION

As the SEEMP is mandatory since 1st January 2013 for all ships engaged in international trade while at the same time there is fierce competition in shipping market, it is almost a necessity to improve the existing solutions and approaches for voyage optimization.

In this paper, a modified Kwon's method has been developed to estimate the ship's added resistance considering the specific ship type. Based on the modified Kwon's method, as well as ship noon reports and sea trial data, a semi-empirical ship operational performance prediction model has been developed to provide accurate ship operational performance prediction under varying drafts, speeds, encounter angles, sea states, fouling effect and engine degradation conditions for each specific ship. Through the two case studies of Suezmax oil tanker A and Aframax oil tanker B, it has been verified that the proposed semi-empirical model is very fast and very reliable on ship operational performance prediction, and may also be used to examine the fouling effect of hull and propeller, and engine degradation trends. Together with a grids system and real-time climatological information, the ships ' various courses can be evaluated according to a number of objectives including maximization of safety, minimization of fuel consumption and voyage time. Finally, by utilizing a decision support tool, the shipmasters as well as shore based route planners may now select the optimum voyage route with minimum fuel consumption considering weather and sea conditions.

8. FUTURE WORK

Since the weather is stochastic, the ship performance simulation needs to be repeated with the weather forecast update frequency, which is normally 4 times a day. As the actual voyage course may not follow the suggested route absolutely, and the latest updated weather forecast may not exactly follow previous forecast, minor changes of the suggested route are expected. However, for a long distance voyage, due to the big uncertainties of long-term weather forecast, bigger changes may be expected by comparing the actual voyage route after arrival with the suggested route before departure.

It would be interesting to examine the applicability of the semi-empirical ship operational performance prediction model to other ship sizes and ship types. Therefore, more case studies will be carried out. Following the development of the ship added resistance model, the next step would be the development of a self-refined ship performance database. The database would be able to store the fuel consumption rate under each sea state, speed, encounter angle, draft and ship conditions (fouling conditions of hull and propeller, and main engine degradation conditions). The feedback from shipmasters would also be recorded to update the ship operational performance records. Using the self-refined ship performance database, the users will be able to extract relevant ship operational performance for a given sea state, speed, draft, encounter angle and ship conditions. With the minimum inputs, the improved accuracy in performance prediction will increase the benefits to be gained from using such systems for energy efficient solutions.

ACKNOWLEDGEMENTS

The study presented in this paper was carried out as part of the research project: Shipping in Changine Climate - funded by UK Research Council (EPSRC Grant no. EP/K039253/1).

REFERENCES

1. Alexandersson, M., 2009. A study of methods to predict added resistance in waves, performed at seaware AB.
2. Avgouleas, K., 2008. Optimal Ship Routing Master thesis. Massachusetts Institute of Technology http://hdl.handle.net/ 1721.1/ 44861.
3. Bunker Index, IFO 380 price in Port Singapore, Rotterdam and Fujairah. Website: ⟨http://www.bunkerindex.com/⟩ (accessed Aug 2014).
4. Christiansen, M., Fagerholt, K., Ronen, D., 2004. Ship routing and scheduling: status and perspectives. Transp. Sci. 38 (1), 1–18.
5. Christiansen, et al., 2013. Ship routing and scheduling in the new millennium. Eur. J. Oper. Res. 228 (3), 325–333.
6. Fagerholt, K., Laporte, G., Norstad, I., 2010. Reducing fuel emissions by optimizing speed on shipping routes. J. Oper. Res. Soc. 61 (3), 523–529.
7. Faltinsen, O.M., Minsaas, K.J., Liapis, N., Skjordal, S., 1980. Prediction of resistance and propulsion of a ship in a seaway. In: Proceedings of the 13th ONR Symposium.
8. Fuji, H., Takahashi, T., 1975. Experimental study on the resistance increase of a ship in regular oblique waves. In: Proceedings of the 14th ITTC, Vol. 4, pp. 351.
9. Gerritsma, J., Beukelman, W., 1972. Analysis of the resistance increase in waves of a fast cargo ship. Int. Shipbuild. Prog. 19 (217), 285–293.
10. Google maps API. Website: ⟨https://developers.google.com/ maps/documentation/javascript/tutorial⟩ (accessed January 2015).
11. Hinnenthal, J., Clauss, G., 2010. Robust pareto-optimum routing of ships utilising deterministic and ensemble weather forecasts. Ships and Offshore Structures. 5, 105–114.
12. Holtrop, J., Mennen, G.G.J., 1982. An approximate power prediction methodInt.Shipbuild. Prog. 29, 166–171.

13. IMO, 2009. Note by International Maritime Organization. Second IMO GHG Study 2009.
14. IMO, 2011. Note by International Maritime Organization. Amendments to the annex of the protocol of 1997 to amend the International Convention for the Prevention Of Pollution from Ships 1972 as Modified n the protocol of 1978 relating thereto (Inclusion of regulations on energy efficiency for ships in MARPOL Annex VI), MEPC 62/24/Add.1, Annex 19.
15. IMO, 2012. Note by International Maritime Organization. 2012 Guidelines for the development of a SEEMP, MEPC 63/23, Annex 9.
16. ITTC, 2011. Note by International Towing Tank Conference. ITTC – Recommended Procedures – Prediction of Power Increase in Irregular Waves from Model Test, 7.5-02-07-02.2.
17. Kuroda, M., Tsujimoto, M., Fujiwara, T., Ohmatsu, S., Takagi, K., 2008. Investigation on components of added resistance in short waves. J. Jpn. Soc. Nav. Archit. Ocean Eng., 171–176.
18. Kwon, Y.J., 2008. Speed loss due to added resistance in wind and waves 3, 14–16Nav. Archit. 3, 14–16.
19. Larsson, E., Simonsen, M.H., 2014. Direct Weather Routing. Chalmers University of Technology, Master's thesis http://studentarbeten.chalmers.se/publication/ 205858-direct-weather-routing.
20. Matulja, D., Sportelli, M., Guedes Soares, C., PRPIĆ-ORŠIĆ, J., 2011. Estimation of Added Resistance of a Ship in Regular Waves Brodogradnja – Shipbuilding.
21. National Oceanic and Atmospheric Administration (NOAA) Website: ⟨www.noaa. gov⟩ (accessed June 2015).
22. Padhy, C.P., Sen, D., Bhaskaran, P.K., 2008. Application of wave model for weather routing of ships in the north indian ocean. Nat. Hazards 44, 373–385.
23. Ronen, D., 1983. Cargo ships routing and scheduling: surveys of models and problems. Eur. J. Oper. Res. 12 (2), 183–192.
24. Ronen, D., 1993. Ship scheduling: the last decade. Eur. J. Oper. Res. 71 (3), 325–333. Salvesen, N., 1978. Added resistance of ships in waves. J. Hydronaut 12 (1), 24–34. Shao, W., 2013. Development of an intelligent tool for energy efficient and low environment impact shipping. University of Strathclyde, Ph.D thesis (PrimoLocal Repository).
25. Wang, X., Teo, C., 2013. Integrated hedging and network planning for container shipping's bunker fuel management. Marit. Econ. Logist., 172–196.

CHAPTER 8

Cell–cell and cell–surface interactions in an illuminated biofilm: Implications for marine sediment stabilization†

Barbara Wigglesworth-Cooksey, Deborah Berglund and Keith E. Cooksey

Montana State University, Department of Microbiology, Bozeman, Montana, 59741, USA.

ABSTRACT

Most wetted surfaces that are illuminated support a population of phototrophs. The marine sediment is no exception and there the major component of the microphytobenthic population is diatoms. These organisms are credited with stabilizing the sediment against physical disturbance by virtue of the extracellular carbohydrate polymers that they elaborate. However, diatoms synthesize and secrete several carbohydrate polymers and it is not certain which of them is involved in the stabilization process. In order to investigate this, we have constructed small glass bead-filled flow through bioreactors to mimic marine sediments. The flow rate through the bioreactors was found to reflect the physical stability of the bead bed. Thus flow rate was measured as a function of diatom growth and the production of three operationally-defined polymers, *i.e.*, those soluble in the medium, those soluble in 0.5 M $NaHCO_3$ at 90 °C and those not soluble in either solvent (matrix polymer). Growth of the diatoms did not change the hydraulic conductivity of the bioreactors. For *Amphora coffeaeformis*, neither did the production of medium-soluble nor $NaHCO_3$-soluble polymers. However, matrix polymer accumulation was directly correlated with a reduction in flow (regression coefficient $R^2 = 0.96$) and stabilization against physical disturbance. Results with species of *Navicula*were not as clear. Both $NaHCO_3$-soluble and matrix polymers were involved in producing the flow reduction. In the same manner we also measured the effect of *Pseudoalteromonas haloplanktis* growth on bead bed hydraulic conductivity and bead bed stability. Growing alone, no effect was found, but in co-culture with a single diatom species, the bacteria reduced the diatom effect on flow through the bioreactors seen earlier, however did not reduce the extent of their growth. Confocal scanning laser microscopy of beads colonized with diatoms alone, or diatoms in co-culture with bacteria, revealed that *P. haloplanktis* was able to inhibit diatom adhesion to the beads. When the bacteria were present there was less matrix polymer evident. We speculate that this interference with diatom metabolic activity was either the result of less matrix polymer synthesis, or its hydrolysis by the bacteria. The results are applicable to mixed species biofilms of this type on surfaces other than sediments.

1. INTRODUCTION

Intertidal mudflats form an important part of an estuarine ecosystem.[1, 2] Their position between the land and the open ocean is responsible for the fact that they bear much of the impact of human activities on the aquatic environment. It is well recognized that the microphytobenthic organisms inhabiting intertidal sediments are responsible for a large part of estuarine productivity[3, 4] and also that of sandy beaches.[5, 6] Related to this is their involvement in the stabilization of estuarine sediment particles against physical disturbance. There is a large body of literature on the physical stabilization of marine sediments[3, 7–18] and most investigators agree that it is the activities of the microphytobenthic organisms, most likely those of the diatoms, that play a major role in the prevention of sediment movement by wave action. The diversity of microorganisms in a nearshore environment is very large but we are not certain which organisms are most active in the stabilization process. Although Rao and Lewin[19] found 352 species of diatoms in the sediments of a small bay with stabilized areas, it is usually members of the genera *Amphora, Navicula* and *Nitzschia* that are most often seen in those areas.[10] It is not possible to exclude the activities of the benthic bacteria from this process,[20] but from a study of the correlation of the chlorophyll *a* content of sediments and their bacterial density, Underwood and Paterson[21] concluded that bacteria did not contribute appreciably to sediment stability. Van Duyl *et al.*[15] studied the coupling between carbohydrate in sediments and the activity of the indigenous bacteria. Their results suggest that a net increase in water-extractable carbohydrate attributable to diatom activity was indirectly coupled to bacterial productivity. Similarly, Murray *et al.*[23] showed that the incorporation of ^{3}H-thymidine into heterotrophic bacterial DNA was coupled to the phototrophic activity of diatoms in a mixed bacterial/diatom biofilm.

The exact manner in which diatoms contribute to the stabilization process is still not entirely agreed upon. There is a consensus however that the extracellular polymers produced by these organisms are responsible *(e.g.,* ref. 3), but it is uncertain which of the several polymers produced by diatoms has the major role. For instance, in most papers where natural (as opposed to model) sediments were sampled, there was a variable correlation between the chlorophyll *a* extracted from the sediment (a biomass indicator) and the amount of colloidal carbohydrate extracted (putative sediment stabilization indicator),[3] depending whether the extracts were made from ridges or runnels of a mudflat area. There were no correlations when the samples were taken from a sandy area of the beach. Underwood and Paterson did find a correlation between the extractable colloidal carbohydrate of sediments and their chlorophyll *a* content, but there was no correlation between the EDTA-extractable capsular carbohydrates from the same sediments and the erosion threshold for the sediment particles. De Winder *et al.*[5] had very similar results to Underwood and Paterson,[21] but Riethmuller *et al.*[22] found that the chlorophyll *a* content of sediments was not correlated with their erodibility and that the results obtained were site specific.

There are possibly several reasons for the differing results obtained by the various groups of workers. Of primary importance is the one mentioned by Dade *et al.*[20] who cautioned that it is not the mere presence of organisms, but their metabolic activities that is driving the stabilization process. This factor is not always considered. Secondarily, the processing of samples is not uniform between laboratories. Whereas some investigators[3, 5] extracted sediments that had been lyophilized, others worked with fresh[22] or frozen[21] sediments. Lastly Decho[13] has observed that EPS is merely an operationally defined substance. It is not a specific compound with defined properties and its ability to be extracted with a particular solvent will vary between organisms and with their physiological state. Further, Decho[13] has observed that it will be important to examine the cohesive properties and viscosities of different types of diatom EPS to understand why some mats may or may not exhibit stabilizing effects on sediment. We have started to address this idea.

In view of the variability of natural sediments and perceived problems with sampling that are related to environmental patchiness, we have designed a completely artificial model system in order to answer questions related to the roles of bacteria and diatoms in sediment stabilization. The work is relevant to the physiological interaction of these organisms in any illuminated biofilm whether it be on the surface of a marine sediment, or on a marine structure such as a ship.

2. MATERIALS AND METHODS

2.1 Growth media

Artificial seawater medium modified to contain 5 mM calcium was used for diatoms (ASP_2,[25]) and marine broth, 2216 [Difco] for bacteria. For some experiments bacteria and diatoms were grown in ASP_2 enriched with 0.2% D-glucose and 0.05% yeast extract [HO medium]. Mixotrophic (10–15 $\mu molm^{-2} s^{-1}$) and heterotrophic growth (darkness) experiments with diatoms were carried out with the addition of 0.5 mM organic substrates, as described previously.[24]

2.2 Organisms

The diatoms were isolated from stabilized patches of sediment. *Amphora coffeaeformis* was isolated from a mangrove swamp drainage[24] and *Navicula* sp.l was isolated from False Bay, San Juan Is., WA. Axenic cultures were made by picking colonies from streaked plates and re-streaking the enrichments until there was no growth of contaminating bacteria in ASP_2 seawater medium enriched with 0.05% D-glucose and 0.02% yeast extract. Bacteria were isolated from the same sediment sample as the diatom *Navicula* sp.l. The sediment particles were washed aseptically with marine medium by decantation. The washed particles were then treated with ultrasound in a low power sonic cleaning bath to remove bacteria attached to sediment particles. This enrichment of attached bacteria was used to inoculate 2216 petri plates for the isolation of experimental organisms. The bacterium used here was identified by its 16-S RNA sequence and fatty acid methyl ester profile as*Pseudoalteromonas haloplanktis* by MIDI labs, Newark, DE and Microcheck Inc., Northfield Falls, VT.

Diatom inocula for experimental purposes were grown at 25 °C and 80–100 $\mu molm^{-2}\ s^{-1}$ without shaking until mid logarithmic phase (3–400 000 cells mL^{-1}). In some instances, diatoms grown for extraction of polymers were kept in 2.8 L

Fernbach flasks containing 250 mL medium. Polymers were extracted from both fresh and lyophilized cells.

Bacteria were grown until late logarithmic phase (absor-bance at 660 nm = 0.5–0.7) with shaking (200 rpm) in 15 mL medium contained in a 125 mL flask with four baffles.

2.3 Analytical methods

Carbohydrates in solution and attached to glass beads were determined by the phenol/sulfuric acid method of Dubois *et al.*[26] and orthophosphate by that in ref. 27. Chlorophyll was determined fluorimetrically.[28] Diatom cells were counted in a hemocytometer. At least 400 cells were counted for each determination to achieve a coefficient of variation (CV) of 10%. Protein was determined by the method of Lowry *et al.*[29] Results are presented as means ± standard deviations. Where these are not shown on graphs it is because they are less than the size of the data point.

2.4 Model sediments

Model sediments in the form of columns of glass beads contained in 10 mL disposable syringes were used as the experimental device. The rationale for this approach was that as the interstitial spaces in the bead bed were occluded by polymers, the flow rate for the column should become reduced. Several methods to produce these bioreactors were investigated. Methods wherein the beads were sterilized *in situ*, wet or dry, proved to be unsatisfactory in that the flow rates of columns were not reproducible. Pre-sterilized columns filled aseptically with a sterilized suspension of beads also gave unacceptable flow rates because the numbers of very small beads (fines) varied from batch to batch. The following method gave control sterile columns with uniform flow rates that did not change appreciably over a period of 12 days. Glass beads with a nominal diameter of 100 μm were boiled in a detergent (Micro, International Products Corp., Burlington, NJ, USA) known to give a very low organic residue. The beads were allowed to fall through a column of high purity water until detergent free and the fines had been removed. This took several cycles of washing and fine removal. Unless the fines were removed, control, un inoculated columns of beads steadily decreased their flow rate with time of incubation. The diameter of the beads produced in this manner was 100 ± 10 μm. A sterile suspension of beads (2 mL beads in 5 mL medium) was pipetted into a sterile 10 mL plastic syringe modified so that a 16 mm translucent cap would fit closely. The bead bed was washed with 12 mL medium and the flow rate of medium through the column measured in triplicate. This volume was needed for the column bed to pack so that flow rates did not vary as a function of the volume that had passed through the column at a later time. The initial flow rate was compared to that after incubation with microorganisms, *i.e.*, each column was its own control. Twenty four columns were prepared for each experiment. This allowed triplicate control columns (uninoculated) and seven tests. Columns of beads were inoculated with 0.5 × 10[6] diatoms in ASP_2 with 0.25 mM calcium and/or 0.5 mL of a bacterial suspension, absorbance 0.5–0.7. The diatoms were prepared in the medium containing lower level of calcium so that a uniform suspension was made[30] which ensured that uniform inoculation of each column. A cell count of the inoculum and a determination of its chlorophyll content allowed the chlorophyll/cell to be calculated.

2.5 Sampling of the columns

At each time point three columns were drained and soluble carbohydrate and phosphate determined. The columns were then re-filled with medium and their flow rates measured. The bead bed was then extracted *in situ* with 0.5 M $NaHCO_3$.[31] This extract was drained and its carbohydrate, together with that left in the extruded beads, was determined. The beads in a second set of three columns were extracted with 90% acetone to determine chlorophyll *a* as an indicator of diatom growth.

2.6 Confocal microscopy

A bead suspension in HO medium contained in a 55 mm petri dish was inoculated with *Navicula* sp.l at zero time, a second dish was inoculated at zero time with diatoms and *Pseudoal teromonas haloplanktis*. A third dish was inoculated with diatoms and after 4 d incubation this was then inoculated with bacteria. Just before microscopic examination, 3 mL of SYBR Green I was added with gentle swirling to stain the bacteria *in situ*. Confocal microscopy was then performed using a Leica microscope, model TCS-SP fitted with a 40 × water-immersible lens. Optical slice images (0.5 μm) were collected at 500–560 nm (green fluorescence) using a 488 nm excitation wavelength from an argon laser. Red fluorescence was collected at 580–680 nm using excitation from a krypton laser (568 nm). Images were analyzed by Adobe Photoshop and Imaris-3D software. Scanning electron microscopy was performed using a JEOL 6100 microscope fitted with a cryostage. Samples were coated with 10 nm of pure gold.

2.7 Staining of microorganisms

Viable and non-viable diatom cells were distinguished from one another by staining with the dye Sytox Green (Molecular Probes Inc., Eugene, OR, USA) in DMSO. Wet mounts were stained for 15 min in 10000 × dilution of the solution provided by the manufacturer (no concentration information for the dye is given by Molecular Probes Inc.) and observed in an epifluorescence microscope using a B2A filter set (excitation filter, 450–490 nm; dichroic mirror, 510 nm; and emission filter, 520 nm; Nikon Instruments Inc.). The final concentration of DMSO in the cell suspension was 5%. Cells with compromised cell membranes stained green, viable cells were not stained. Bacteria were also stained with a 1/200 dilution in growth medium of the stock solution of SYBR Green I supplied by Molecular Probes Inc. At this dilution, no rinsing of the stained cells was needed and the background did not fluoresce. They were observed with a B2A epifluorescence filter set or in the confocal microscope.

Diatom adhesive plaques or footpads were stained either with 0.1% Acridine Orange (AO) in water for 15 min and washed with water, or Concanavalin A conjugated to fluor escein isothiocyanate and washed with growth medium. Footpads stained orange with AO and yellow–green with Con A when observed using the B2A filter cube

2.8 Production of footpads

Diatom cells were grown on the surfaces of microscope slide cover glasses, microscope slides or in small four-place bioreactor/culture slides (Becton-Dickenson, Franklin Lakes, NJ, USA) for 2 d and then treated with 10 mM EGTA, pH 7.8 or 0.2 M EDTA, pH 7.25 for 25 min.[32] The cells were washed with medium and observed with dark phase optics or after staining.

2.9 Extraction of polymers

(i) To determine the effect of the state of the diatom cells on the relative amount and quality of the polymers extracted, we extracted both fresh and lyophilized cells using 1.5 M NaCl as the extractant.[33]

(ii) Water soluble carbohydrate polymers were considered to be soluble in the growth medium and were measured in the column effluent produced by draining the column when it was sampled. Polymers soluble in 0.5 M $NaHCO_3$ at 90 °C were extracted *in situ* by adding 5 mL of the solution to the column and then draining one bed volume of liquid from the column. The fact that the pH of the effluent changed from pH 7.8 to about 8.5 showed that the column liquid was fully exchanged. The columns were incubated for 1 h at 90 °C and then drained. The remaining carbohydrate polymers were determined directly on the glass beads. These polymers can be seen microscopically (dark phase) after bicarbonate extraction as amorphous clear structures surrounding the cells. The matrix fraction could be contaminated by intracellular carbohydrate-reacting materials not removed by the previous treatment. Contamination of one fraction with another because of the volume of liquid retained by the bead bed was corrected by using a knowledge of the hold up volume.

3. RESULTS

3.1 Growth of diatoms

We have shown previously[24] that *Amphora coffeaeformis* grows in ASP_2 medium with a generation time of 12 h at 25–28 °C. The organism can grow heterotrophically (dark growth) and mixotrophically (stimulation of light-limited growth) on D-glucose, D-fructose, L-glutamate and yeast extract. *Navicula* sp.l, on the other hand, has an autotrophic generation time of 32 h, is heterotrophic only on 0.05% yeast extract and mixotrophic on 0.5 mM D-glucose or L-glutamate.

3.2 Extraction of putative diatom polymers

The extracts of fresh and lyophilized diatoms were of very different composition. Table 1 shows that a much larger amount of both protein and carbohydrate reacting material was extracted from the lyophilized cells than from the cells in a fresh state. When samples of these cells were stained with Sytox Green and examined microscopically, the percentage with compromised membranes was found to be 11% (n = 633) for the fresh cells and 100% (n = 300) for the lyophilized cells. Microscopic examination of footpads, shown in Fig. 1, indicated that they were soluble in hot (90 °C) bicarbonate solution.

Table 1. Analysis of extracts from fresh and lyophilized diatom cells

	Amphora[a]		*Navicula*[a]	
Analysis	Fresh	Lyophilized	Fresh	Lyophilized
Carbohydrate	3.6	68.8	1.6	82.7
Protein	24.0	135.0	1.4	168.3

[a]Quantities are given in μg mg^{-1}dry wt. of cells.

3.3 Experiments with glass bead-filled columns

(i) Diatoms alone. Fig. 2 shows that the growth of *A. coffeaeformis* on beads leads to their stabilization against a mechanical force, in this case gravity. The image was made with the bioreactor turned at 60° to the horizontal plane. In each case, the control uninoculated beads adopted the horizontal position (avalanched), whereas those with diatoms attached did not move.

Fig. 3 shows the influence of the growth of *Amphora* on the reduction in flow rate of the packed bead beds. This reduction was not evident until the phosphate in the medium had been reduced from 24 μM to about 0.5 μM. Since cells ceased increasing in chlorophyll content after 120 h, and therefore most probably in number, it is not likely that the cells themselves reduced the flow through the bioreactor. Fig. 4 indicates that reduction in flow was also not correlated with production of bicarbonate-soluble carbohydrate polymers. Neither did the production polymers soluble in the medium influence flow rate (data not shown). However, flow reduction was correlated R^2= 0.96) with matrix polymer synthesis (Fig. 5). The results for *Navicula* sp.l appeared similar (Fig. 6 and 7), but not identical to those for *Amphora*. Again, carbohydrate polymers soluble in the medium had no influence on the flow through the bioreactors (Fig. 7), but there was no obvious correlation with any of the carbohydrate fractions insoluble in the medium. Both matrix and bicarbonate-soluble polymers increased more rapidly after the medium became phosphate limited (120 h). In contrast to *Amphora*, the bicarbonate-soluble polymer rather than the matrix polymer fraction increased most after the phosphate had been depleted. Even when these three polymers were summed (Fig. 8), the correlation between this parameter and flow reduction was not strong.

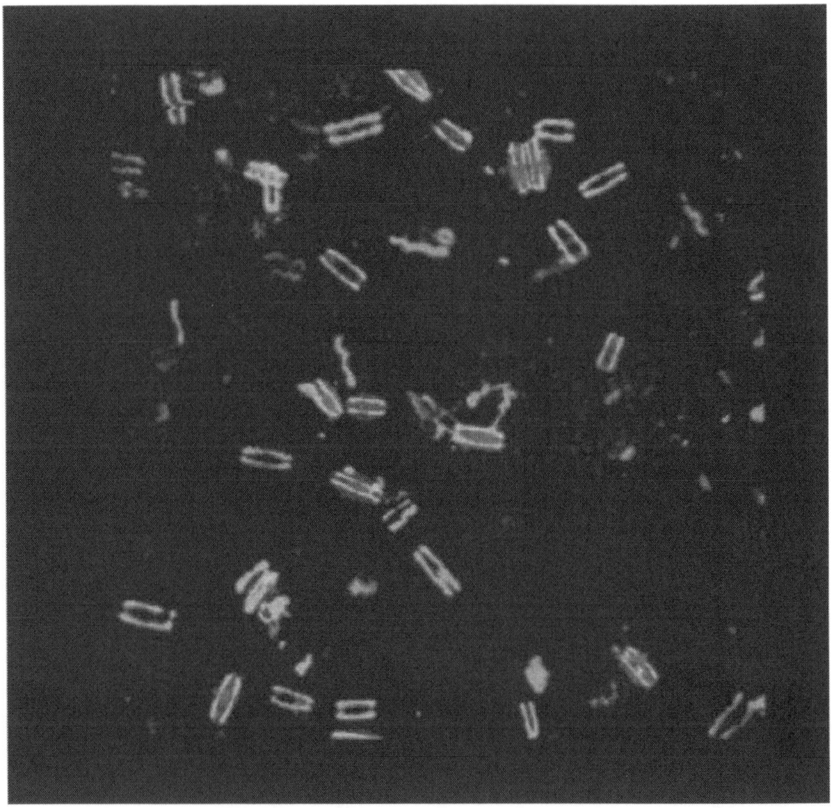

Figure 1. Substratum-attached material ("footpads") remaining after treatment of *A. coffeaeformis* with EGTA. The footpads are soluble in hot bicarbonate solution and are stained here with Concanavalin A conjugated to fluorescein isothiocyanate (100 μgmL^{-1}, Sigma Chemical Co.). Red autofluorescence of the diatom chloroplast is seen above the footpads when diatoms were not removed from the substratum.

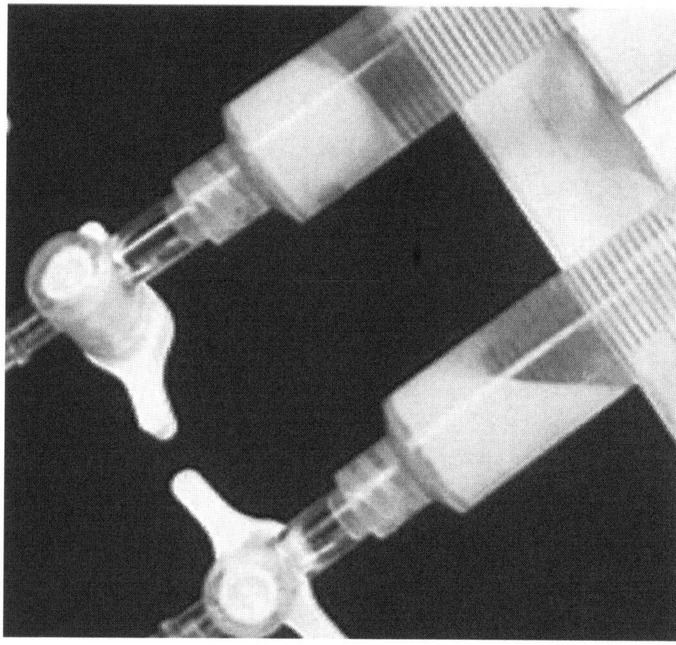

Figure 2. Bioreactors containing glass beads (100 μm). Upper bioreactor was inoculated with *A. coffeaeformis* 10 d previously. At this time diatoms were no longer growing and matrix polymer formation was well developed. Lower bioreactor is uninoculated control. Note avalanching of beads in the uninoculated control.

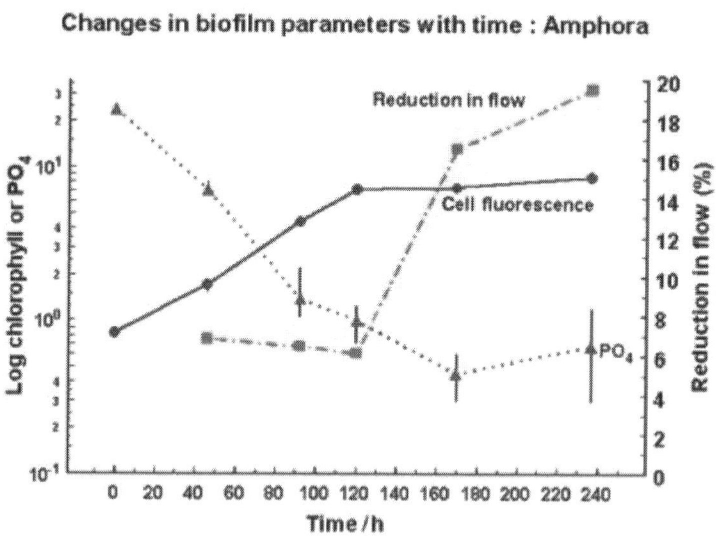

Figure 3. Changes in biofilm parameters with time: *A. coffeaeformis*.

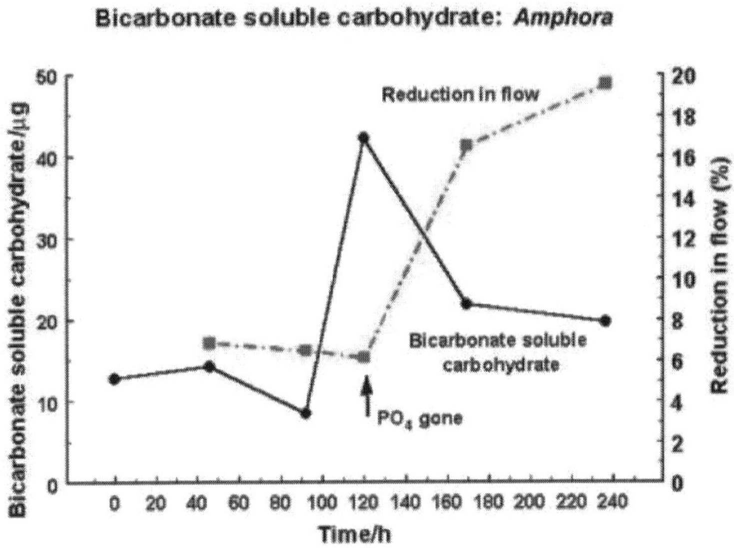

Figure 4. Extracellular carbohydrate polymer that is soluble in bicarbonate solution: *A. coffeaeformis*.

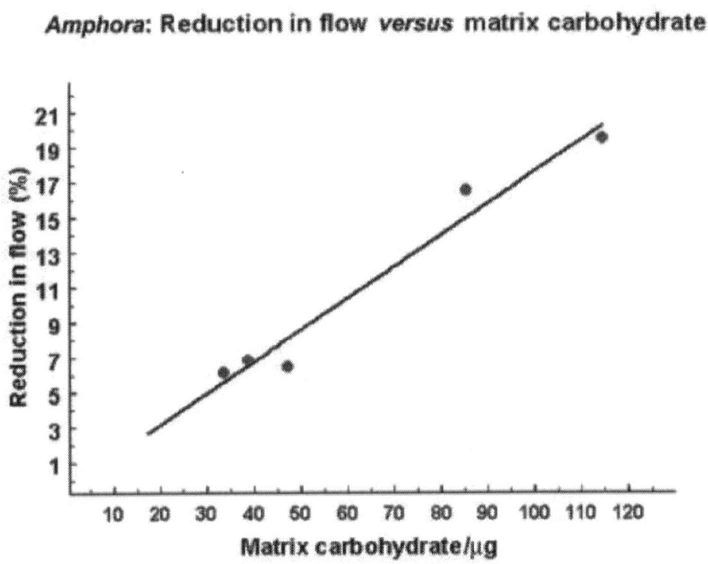

Figure 5. Reduction in hydraulic conductivity (flow) of bioreactors during growth of *A. coffeaeformis* and concomitant production of matrix polymer. Regression coefficient, $R^2 = 0.96$. As seen in Fig. 3, this curve represents matrix polymer formation over the same period of 45–240 h.

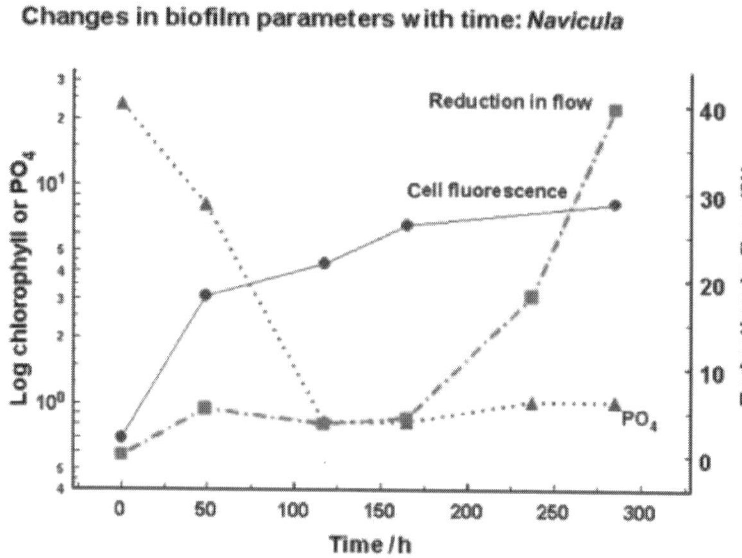

Figure 6. Changes in biofilm parameters with time: *Navicula* sp. 1.

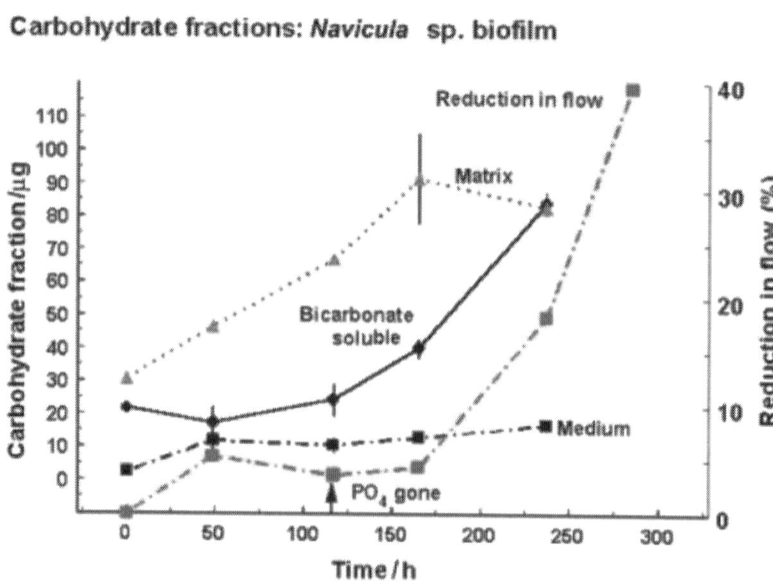

Figure 7. Production of extracellular carbohydrate polymers with time: *Navicula* sp.l.

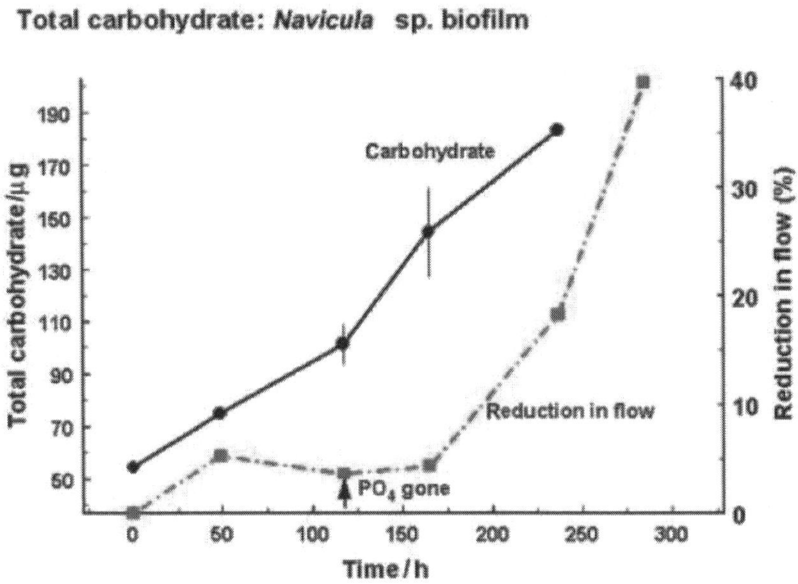

Figure 8. Individual carbohydrate polymer fractions summed to produce a total carbohydrate value: *Navicula* sp.l.

(ii) Experiments with diatoms and bacteria in mixed cultures. The influence of bacteria from sediments on the process of bead bed stabilization was investigated using culture of diatoms alone and diatoms in admixture with the bacterium *Pseudoal teromonas haloplanktis*. We found that the presence of bacteria in a diatom culture gave different results than those obtained with diatoms alone (Fig. 9). The ability of the diatoms to reduce the hydraulic conductivity of the bioreactor bead bed was inhibited by the presence of the bacteria. However flow rates with bacteria alone were not statistically different from those in the uninoculated controls (compare columns 1 and 2, Fig. 9). The flow rates through bioreactors inoculated with *Amphora* and *Navicula* alone or a mixture of both organisms were not statistically different (compare columns 5, 6 and 7, Fig. 9). However, the flow rates with bacteria and diatoms in co-culture were greater than those obtained with a single diatom species (compare columns 3 and 6; 4 and 7, Fig. 9), but not when diatoms of both species were present (compare columns 5 and 8, Fig. 9). The relative fluorescence values for chlorophyll extracted from the bioreactors inoculated with a single diatom culture and those inoculated with diatoms plus bacteria were not different. For example, the values for *Amphora* alone and *Amphora* with *Pseudoalteromonas* were 8.65 and 8.75; the equivalent figures for Navicula were 6.45 and 6.96.

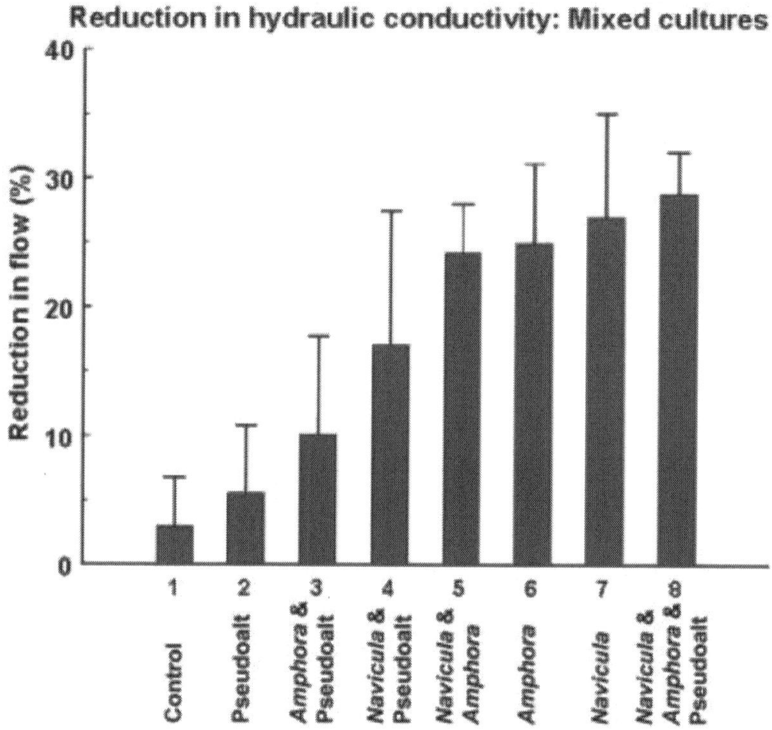

Figure 9. Reduction in hydraulic conductivity of bioreactors inoculated with mixed cultures. Bars represent means ± SD ($n = 3$).

In an effort to elucidate the reasons for these results we examined glass beads colonized with diatoms alone and those colonized with mixtures of bacteria and diatoms using both SEM and CSLM. Fig. 10 shows a typical biofilm of *Navicula* and *Pseudoalteromonas* growing on beads. The micrograph does not provide clear clues to explaining the bacterial–diatom interaction, but it does demonstrate that the bacteria have preferentially colonized the bead surface. Fig. 11 supports this idea. Panels 11 (a) and ll(d) show complete coverage of the beads by*Navicula* sp.l whereas panels ll(b) and ll(e) and ll(c) and ll(f) show that in the presence of bacteria, diatom colonization is inhibited. Bacteria in these images are green (SYBR Green 1 staining) and diatoms are red (chlorophyll autofluorescence). Note that some of the bacteria are seen as streaks showing that some of them were motile. This indicates that the stain did not kill them. Motile diatoms also showed as streaks. Such cells were seen only in the presence of bacteria [compare image 11(a) with 11(b) and (c)]. These interpretations were supported by results shown in Fig. 12(a) and 12(b). Images were obtained by collecting optical "slices" of the bead at its center so that only the bead circumference can be seen. The differential colonization effect between diatoms and diatoms with bacteria is obvious. In admixture, bacteria colonize the beads preferentially.

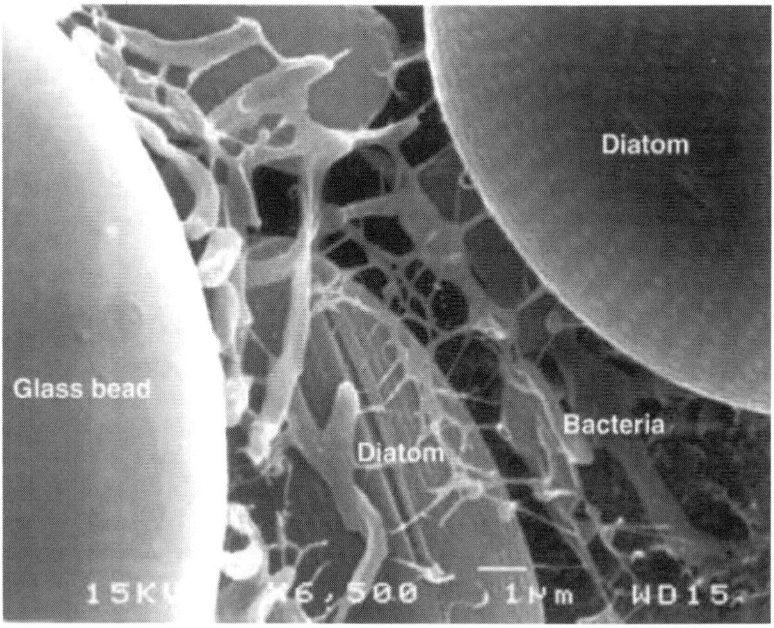

Figure 10. Scanning electron micrograph of a mixed culture of *Navicula* sp. 1 and *P. haloplanktis* on 100 μm glass beads. (Equipment: JEOL 6100 instrument fitted with a cryostage, sample was coated with 10 nm gold).

4. DISCUSSION

There is general agreement that the metabolic activities of the benthic diatom community influence the physical stability of marine sediments and that extracellular carbohydrate polymers synthesized by diatoms play an important role in the process.[3, 5, 22] However there is disagreement concerning the details of the stabilization process and which extracellular polymers are involved, *e.g.*, ref. 22. We suggest that much of the disagreement can be traced to differing methodologies between laboratories and what appears to be a tacit assumption that all diatom species behave similarly. If they do not, field investigations where speciations of the biofilm populations differ will not agree. For instance, extracellular carbohydrate polymers may well be extracted from one species of diatom with a certain efficiency and a differing efficiency from another species. Further complication is that diatoms of a single species secrete polymers differing in structure, depending on the growth phase.[34, 40] A sediment will contain many species in various stages of the cell cycle. The fact that most diatoms secrete increased levels of extracellular polymer in the transition to stationary phase is however agreed, as is the influence of phosphate limitation on this process. Phosphate limitation stimulates "overflow metabolism" which results in increased polymer secretion (ref. 17 and references therein). Our argument is applicable to any illuminated biofilm, not just those on sediments.

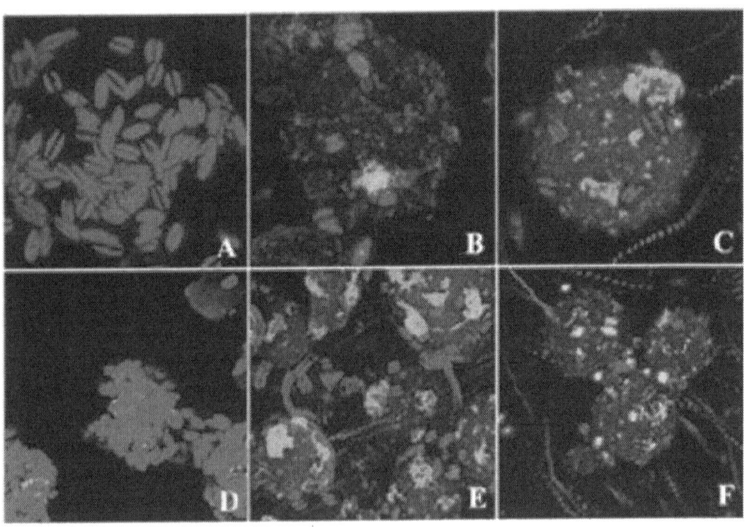

Figure 11. Confocal microscope images of colonized glass beads. Diatoms (*Navicula* sp. 1) are seen by the red fluorescence of their chloroplast. Bacteria (*P. haloplanktis*) are stained with SYBR Green 1. (a) and (d) Diatoms alone; (b) and (e) *P. haloplanktis* added to the diatom culture after 4 d; (c) and (f) *P. haloplanktis* added to diatom culture at time zero. Upper panel magnification was 10006 × , lower panel magnification was 23006 ×.

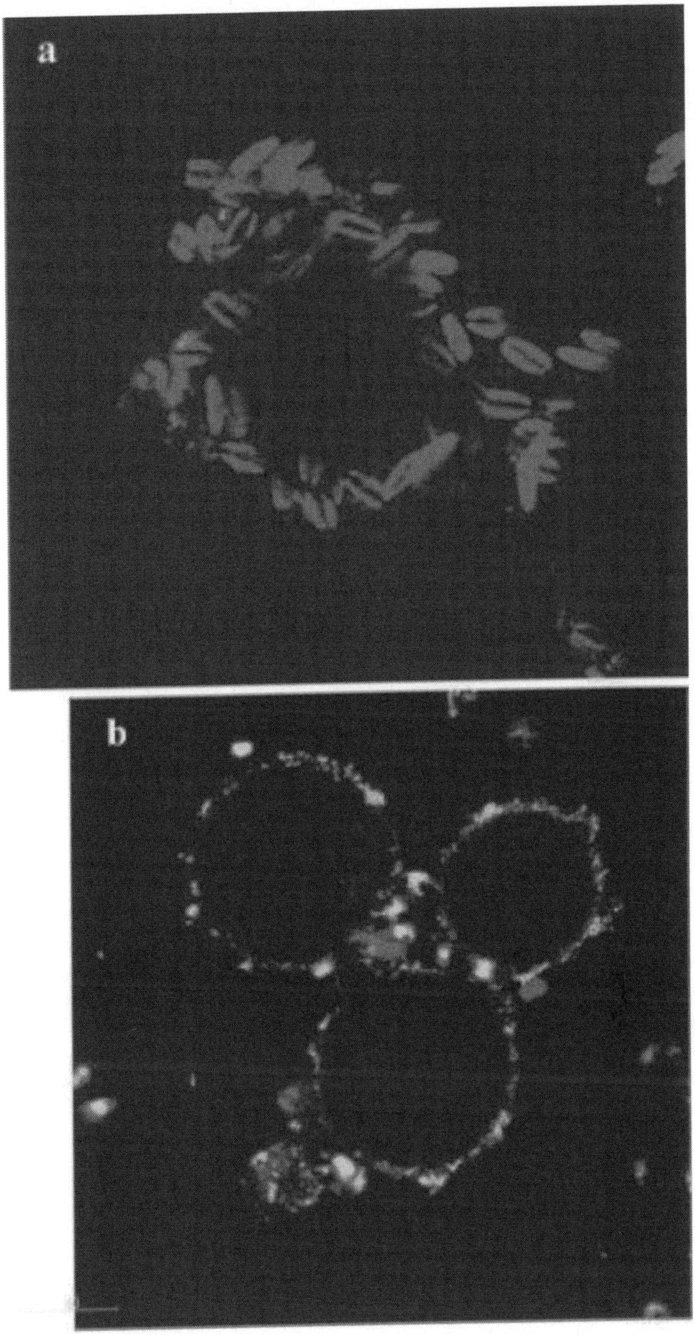

Figure 12. Confocal microscope images of colonized glass beads, (a) *Navicula* sp.l; (b) *Navicula* sp.l and *P. haloplanktis*. Both (a) and (b) are optical "slices" through the centre of the beads, thus cells are seen on the periphery of the beads.

Our results with lyophilized and fresh cells (Table 1) support the idea that the lyophilization process causes cell lysis and thus makes the intracellular polymers available to solvents. Cells examined microscopically before and after lyophilization showed very little difference with light or dark phase optics, but were clearly different when viewed with epifluorescence optics after being stained with a cell membrane impermeant dye. Both Underwood *et al.*[33] and Staats *et al.*[36] discuss the possibility that intracellular polymers, notably chrysolaminarin–a β 1–3 glucan, were extracted from lyophilized cells using solutions of EDTA, NaCl or water alone. Both groups conclude from indirect evidence that intracellular polymers are not extracted. Neither group however monitored membrane integrity. Underwood *et al.*[33] did show that the yield of colloidal polymer from sediments was increased two-fold after lyophilization. Most investigators routinely lyophilize sediment samples to preserve the integrity. In the light of our results we suggest that perhaps the extraction process should be re-evaluated, de Brouwer *et al.*[16] have made an unpublished observation that lyophiliza tion causes cell lysis in diatoms.

We have measured three operationally defined extracellular carbohydrate polymers. These are: (i) the polymer easily soluble in artificial seawater (24ppt), (ii) the polymer secreted by stationary cells before they adhere permanently and which is not soluble in saline medium but is soluble in hot bicarbonate solution; and (iii) the polymer remaining after bicarbonate soluble polymers have been removed. Fraction (i) will contain polymer secreted as a result of motility; (ii) represents the water insoluble, bicarbonate soluble polymers referred to as WIBS by Wustman *et al.*[31]; and (iii) represents the matrix polymer visibly responsible for biofilm architecture and most likely the product of "overflow metabolism".[17] There is a potential for this fraction to contain also any intracellular carbohydrate-reacting material not extracted earlier. The WIBS fraction has been investigated by Wustman *et al.*[31] in detail. The sugar constituents of this fraction from *Amphora coffeaeformis* and*Achnanthes longipes* did not differ appreciably in either sugar content or their amounts, as determined by GC-MS analysis. However, mechanically isolated polymers from *Amphora* differed considerably in composition from the WIBS fraction of the same organism and from the mechanically isolated stalks of *Achnanthes*. Although we have no comparable information on *Navicula*, it seems reasonable to suggest that extracellular polymers of this organism could differ from those *of Amphora*. This is suggested by the time courses of accumulation of polymers in this fraction (*cf.* Fig. 4 and 7).

The mere presence of diatom cells on the surfaces of the glass beads in the bioreactors did not interfere with flow through the bead bed or influence its physical stability. With both *Amphora* and *Navicula*, the reduction in flow and stabilization of the bead bed (Fig. 2) did not occur until phosphate concentration in the medium fell below an assimilable level and growth, as measured by chlorophyll fluorescence, ceased. The fact that growth measured by this parameter ceased when phosphate became limiting allows confidence in the use of chlorophyll as a biomass level indicator. It can be seen from Fig. 2, 3 and 5 that the stabilization of the bead bed, the reduction of its hydraulic conductivity and the production of matrix polymer occurred in the same time-frame.

Our results with glass bead-filled bioreactors do not support the involvement of growth medium-soluble polymers in sediment stability because this polymer fraction did not change throughout the course of the diatom growth. Its amount was not correlated with the reduction in hydraulic conductivity of the bead bed. In this respect, our results agree with those of Staats *et al.*[36] and van Duyl *et al.*,[37] but not those of Underwood *et al.*[33] or Patterson.[35] Staats *et al.*[36] used both axenic cultures of *Cylindrotheca closterium* and mudflat samples to measure exopolysaccharide secretion in various conditions. Their results did not support the involvement of the motility polymer in the accumulation of exopolysaccharide in the light. The results of Underwood *et al.*[33] support their involvement. To our knowledge, no one working in the field of microbial sediment stabilization has suggested that a bicarbonate-soluble diatom exopolymer component contributes to the stabilization process. Our choice of this fraction was because Wustman *et al.*[31] stated that "...the majority of the adhesive polymers were contained in this fraction..." In *A. coffeaeformis* we found that this fraction did not contribute to sediment stabilization as measured by our flow method. It is important to realize that in the early stages of growth, this fraction probably contained internal carbohydrates in addition to those from the extracellular environment, but that this is less likely in later growth stages once the cells had become phosphate limited. Staats *et al.*[11] showed that intracellular carbohydrate polymers did not increase during phosphate limitation. The fraction responsible for stabilization appears to be the matrix polymer. This polymer would appear in the total carbohydrate of Underwood *et al.*[33] Our results suggest that its accumulation is controlled by the level of phosphate in the medium. This type of response in diatoms is well known (refs. 17, 38, 39 and references therein).

The results of identical experiments with *Navicula* sp.l indicate that it is unwise to generalize based on results with a single organism. Reduction in flow occurred when phosphate concentration limited growth, but the WIBS fraction and the matrix polymer were produced throughout the growth period, as well as in the stationary phase, although the WIBS production did increase at phosphate limitation (Fig. 7). It can be seen from Fig. 8 that indeed the production of total extracellular carbohydrate polymers was correlated with reduction in flow through the bioreactors but that the correlation was not as clear cut as with *Amphora*. Smith and Underwood[40] using pyrolysis mass spectrometry, also showed that diatom EPS differs from species to species and, depending on the phase of growth, within species.

In view of the known interactions between diatoms and bacteria,[23] our initial hypothesis concerning mixed species biofilms and sediment stabilization was that it was likely that there would be a positive or may be a synergistic interaction leading to an enhanced sediment stabilization. Our results do not support this hypothesis (Fig. 9). The presence of *P. haloplanktis* and diatoms actually increased the flow through the bioreactors over that found with diatoms alone. This was true for either *Amphora* or *Navicula* with bacteria, but not true when *both*diatom species and bacteria were combined. It was considered possible that the results with single diatom species and bacteria could arise from an inhibition of diatom growth and/or metabolism by the bacteria. However, as seen by chlorophyll *a* analysis, we found that there was no inhibition of diatom growth. Note that under the conditions of these experiments, the diatoms and the bacteria could utilize the organic substrates in the medium. It is not clear why, when two diatom species and the bacteria are grown together, the flow rate is not different from that obtained in the absence of the bacteria. Van Duyl *et al.*[15, 37] showed that in natural sediments, the activities of heterotrophic bacteria were coupled to the production by diatoms of extracellular carbohydrate polymers. This activity, which they suggest was mediated by β-glucosidase, was dependent on the presence of water-soluble, but not water-insoluble, carbohydrate polymers. These authors also suggest that the utilization of the extracellular polymers by bacteria and the concomitant production of bacterial EPS could also contribute to sediment stabilization. Our results do not support this, but another speculation by the same authors, *i.e.*, rapid reduction of extracellular carbohydrate pools by bacteria, could be a destabilizing force in already stable sediments. For instance, Fig. 10 shows very little extracellular polymer in the interstitial spaces of the beads. In addition, the CLS micrographs (Fig. 11 and 12) show distinctly the differential colonization pattern for diatoms alone and diatoms in the presence of bacteria. Preliminary experiments have shown that marine sediment bacteria produce in the growth media materials that inhibit colonization of surfaces by diatoms. Gawne *et al.*[41] have shown that the story concerning bacterial influences on colonization patterns of *Achnanthes longipes* is highly complex. Elucidation of the means by which *P. halopklanktis* can control diatom colonization patterns is the focus of our continuing study. The implications for this work are not limited to the marine environment. Battin and Sengschmitt[42] indicate that similar microbial phenomena can operate in rivers.

5. CONCLUSION

For the diatom and bacterial species we have studied, we conclude that: (i) Diatom matrix extracellular carbohydrate polymer is largely responsible for sediment stabilization and thus biofilm architecture; (ii) soluble diatom motility polymer plays no part in the sediment stabilization process; (iii) possibly because of their lytic activities on exopolymers elaborated by diatoms, bacteria are more likely to inhibit than enhance diatom driven sediment stability; and (iv) it is unreasonable to make general predictions based on the results obtained with one, or perhaps several, organisms.

NOTE

†Presented during the ACS Division of Geochemistry symposium 'Biogeochemical Consequences of Dynamic Interactions Between Benthic Fauna, Microbes and Aquatic Sediments', San Diego, April 2001.

ACKNOWLEDGEMENTS

We thank the US Office of Naval Research for the support of our work and for their support of the American Chemical Society Symposium on "Biogeochemistry of interactions between infauna, microorganisms and aquatic sediments" where this paper was initially presented. We also acknowledge the support of the American Chemical Society Petroleum Fund for the support of the symposium. Scott Brady is thanked for his technical assistance.

REFERENCES

1. de Jonge VN: Estuarine Coastal Shelf Sci. 1985, 21: 607
2. Admiraal W: Progr Phycol Ser. 1984, 3: 269-
3. Blanchard GF, Paterson DM, Stal LJ, Richard P, Galois R, Huet V, Kelley J, Honeywill C, Brouwer J, Dyer K, Christie M, Seguignes M: Cont Shelf Res. 2000, 20: 1243
4. Underwood GJC, Provot L: Eur J Phycol. 2000, 35: 173
5. de Winder B, Staats N, Stal LJ, Paterson DM: J Sea Res. 1999, 42: 131
6. Yallop ML, Paterson DM, Wellsbury P: MicrobEcol. 2000, 39: 116-
7. Grant J, Bathman UV, Mills EL: Estuarine Coastal Shelf Sci. 1986, 23: 225
8. Fraenkel L, Mead DJ: J Sediment Petrol. 1973, 43: 1090-
9. Johnson RG: J Mar Res. 1974, 32: 313-
10. Madsen KN, Nilsson P, Sunback K: J Exp Mar Biol Ecol. 1993, 170: 159
11. Holland AF, Zingmark RG, Dean JM: Mar Biol. 1974, 27: 191
12. Sutherland TF, Grant J, Amos CL: LimnolOceanogr. 1998, 43: 65-
13. Decho AW: Cont Shelf Res. 2000, 20: 125
14. Tolhurst TJ, Riethmuller R, Paterson DM: Cont Shelf Res. 2000, 20: 1317
15. van Duyl FC, de Winder B, Kop AJ, Wollenzien U: Mar Ecol Progr Ser. 1999, 191: 19de Brouwer JFC, Bjelic S, de Deckere EMGT, Stal LJ: Cont Shelf Res. 2000, 20: 1159
16. Staats N, Stal LJ, Mur LR: J Exp Mar Biol Ecol. 2000, 249: 13

17. Cooksey KE, Wigglesworth-Cooksey B: Diatoms in Biofilms. Encyclopedia of Environmental Microbiology. Edited by: Flemming H-C. Wiley, New York

18. Rao VNR, Lewin JC: Syesis. 1976, 9: 173-

19. Dade WB, Davis JD, Nichols PD, Nowell ARM, Thisle D, Trexler MB, White DC: Geomicrobiol J. 1990, 8: 1

20. Underwood GJC, Paterson DM: J Mar Biol Assoc UK. 1993, 73: 871

21. Riethmuller R, Heineke M, Kuhl H, Keuker-Rudiger R: Cont Shelf Res. 2000, 20: 1351

22. Murray RE, Cooksey KE, Priscu JC: Appl Env Microbiol. 1986, 52: 1177-

23. Cooksey KE, Chansang H: J Phycol. 1976, 12: 455

24. Provasoli L, McLaughlin JJA, Droop M: Arch Mikrobiol. 1957, 25: 392

25. Dubois M, Gilles KA, Hamilton JK, Reber PA, Smith F: Anal Chem. 1956, 28: 350

26. . A Practical Handbook of Seawater Analysis. Edited by: Strickland JDH, Parsons TR. 1972, Fish Res. Bd., Canada, pp 310-2

27. Yentch CS, Menzel DW: Deep Sea Res. 1963, 10: 653-

28. Lowry OH, Rosebrough NH, Farr AL, Randall RL: J Biol Chem. 1951, 193: 265-

29. Cooksey KE: Appl Env Microbiol. 1981, 41: 1378-

30. Wustman BA, Gretz MR, Hoagland KD: PlantPhysiol. 1997, 113: 1059-

31. Cooksey KE, Cooksey B: Adhesion of fouling diatoms to surfaces : Some biochemistry. Algal Biofouling. Edited by: Evans LV, Hoagland KD. 1986, Elsevier, Amsterdam, 41-53.

32. Underwood GJC, Paterson DM, Parkes RJ: Limnol Oceanogr. 1995, 40: 1243-

33. Smith DJ, Underwood GJC: Limnol Oceanogr. 1998, 43: 1578-

34. Paterson DM: Limnol Oceanogr. 1989, 34: 223-

35. Staats N, Stal LJ, de Winter B, Mur LR: Mar Ecol Prog Ser. 2000, 193: 261-

36. van Duyl FC, de Winder B, Kop AJ, Wollenzien U: Cont Shelf Res. 2000, 20: 1335

37. Guerrini F, Cangrini M, Boni L, Trost P, Pistocchi R: J Phycol. 2000, 36: 882-

38. Alcoverro T, Conte E, Mazella L: J Phycol. 2000, 36: 1087

39. Smith DJ, Underwood GJC: J Phycol. 2000, 36: 321

40. Gawne B, Wang Y, Hoagland KD, Gretz MR: Biofouling. 1998, 13: 137

41. Battin TJ, Sengschmitt D: Microbial Ecol. 1999, 37: 185

CHAPTER 9

The interaction of marine fouling organisms with topography of varied scale and geometry: a review

Felicia Wong Yen Myan1*, James Walker1 and Odette Paramor2

1Department of Civil Engineering, The University of Nottingham Ningbo China, Room 412, Science and Engineering Building, 199, Taikang East Road, Ningbo 315100, China

ABSTRACT

Many studies have examined the effects of surface topography on the settlement behaviour of marine organisms and this article reviews these investigations with more emphasis on the effects of topography scale. It has been observed that macro topographies (1-100 mm) are generally favoured by marine fouling taxa and are unsuitable for antifouling applications. This is because macro topographies are usually large enough to fit fouling organisms and provide refuge from dangers in the marine environment. Micro topographies had only limited success at reducing fouling from a wide range of marine taxa. The antifouling performance of micro topographies (1 to ≤1000 μm) is dependent on the properties of topography features in terms of symmetry, isotropy, width, length, height/depth, separation distance and average roughness. In terms of the antifouling performance of micro topography, topography geometry may only be of secondary importance in comparison to the size of features itself. It is also noted that hydrodynamic stresses also contribute to the settlement trends of foulers on textured surfaces. Future studies on antifouling topographies should be directed to hierarchical topographies because the mixed topography scales might potentially reduce fouling by both micro and macro organisms. Patterned nano-topographies (1- ≤1000 nm) should also be explored because the antifouling mechanisms of these topographies are not yet clear.

KEYWORDS

Marine biofouling; Antifouling; Settlement behaviour; Aspect ratio; Nanostructures; Micro/Macro; Topography; Tortuosity

1 REVIEW

1.1 Introduction

Solid surfaces that spend long periods of time in aquatic environments are susceptible to the accumulation of marine fouling organisms and this phenomenon is known as marine biofouling. This is a natural process which can have significant economic impact on maritime industries. For example, the fouling of ship hulls increases drag and friction which in turn increases fuel consumption by 20.4% or US$2.3 million per ship every year [1]. In the aquaculture industry, approximately US$1.5-3 billion is spent a year on antifoulants and on repair or maintenance fees to manage this phenomenon [2]. Significant effort is focused on the search for an effective, environmentally friendly, antifouling solution and there has yet to be any development that can rival the efficacy of tributyltin (TBT) self-polishing copolymers. Fortunately, these were banned in 2008 because of their detrimental effects on marine life, which included severe deformities in shellfishes and the accumulation of tin in other species [3, 4].

Current research to control biofouling is focused on producing non-toxic, effective antifouling systems. Many studies have identified that surface chemistry can affect the settlement trends of marine organisms. These studies resulted in derivations of non-toxic foul-release coatings (FRCs); some of which are commercially available especially for the shipping industry [5–9]. FRCs are made of materials with chemical properties that reduce the adhesion or promote the release of potentially settling organisms [10]. However, these coatings are only effective above a certain level of hydrodynamic shear and still have difficulty preventing the formation of biofilms and the growth of microalgae. Several reviews on foul-release coatings were published in recent years and will not be discussed in this review [10–13].

Another non-toxic antifouling approach involves the development of foul-resistant topographies [14–17]. Some of these topography designs were inspired by naturally occurring marine topographies from organisms that are able to prevent fouling on themselves despite spending most of their lives underwater. Examples of these natural topographies include the micro sized riblets on shark's skin and the ridge-pore covered skin of pilot whales [18, 19]. Artificially derived geometries of varying scales (from macro to nano) such as meshes, pillars, grooves, bumps and holes were also investigated to determine their effect on the settlement behaviour of marine organisms [20–22]. Studies have shown that micro topography has antifouling properties but topography of this scale alone only reduced the settlement of a few marine species at a time [23,24]. Several studies have mentioned that the mixed scale on hierarchical structures might be able to discourage a wider range of fouling organisms from settling concurrently [23, 25, 26]. Since not much is known about this topography configuration, future studies are necessary to determine if it would be more effective than current foul-resistant topographies.

Both FRCs and fouling resistant topographies have had a degree of success at reducing or discouraging fouling, but each approach has its limitations. Some studies have suggested that an optimum antifouling system would require manipulating multiple physico-chemical properties of a surface [25, 27, 28]. It is noted that foul-resistant topographies are made of materials with chemical properties that contribute to the overall antifouling performance of the topography. Several studies explored quantitative models that considered the combined effects of surface topography and chemistry/energy towards the settlement trends of *Ulva* spores [29, 30]. A recent study applied a derivation of these models and determined that micro topographies made of cross-linked hydrogels would exhibit better antifouling properties than the same patterns developed with the silicone elastomer, Polydimethylsiloxane (PDMSe) [31]. These are among the many studies that acknowledged the necessity to consider the interaction between multiple physico-chemical properties for an effective antifouling solution. However, for this review, the chemical aspects of foul-resistant topographies are not included. The aims of this review are to focus on certain physical aspects of topography that contribute to its antifouling properties. This article compiles and summarises the influence of topography scale (i.e. width, length, height/depth, space apart or measured roughness) in relation to theories (e.g. attachment point theory) or derived empirical models (Engineered Roughness Index (ERI)) that were developed to better understand the settlement trends of fouling organisms.

1.2 Methods

This review examined 52 published journal articles from 1988 to 2013 and one PhD thesis. These articles investigated the effects of several physical aspects of surface topography with more focus on topography scale and geometry on the settlement behaviour of marine fouling organisms (Additional file 1: Table S1). Topography features of the same shape or approximate profiles were re-categorised from the original literature and renamed for standardisation purposes. For example, several studies tested a range of macro and micro sized V-shaped trenches or slots [20, 32–36]. Although the shape of the tested geometry was the same, a couple of these studies referred to these slots as riblets [20, 32]; while the rest named them V-shaped crevices or grooves. Therefore, to compile findings in Additional file 2: Table S2 and Additional file 3: Table S3, the name of these trenches were standardised by naming them V-shaped grooves. Additional file 4: Table S4 provides rough illustrations of the geometries to better elucidate the types of topography and their categorisations with respect to discussions in this article.

The effects of topographies on tested taxa were summarised in Additional file 2: Table S2 and Additional file 3: Table S3 according to taxa, topography scale and topographic geometry. The defined range of macro, micro and nano topography is 1-10 mm, 1 to ≤1000 μm and 1- ≤ 1000 nm respectively. Additionally, studies that investigated the antifouling properties of hierarchical structures are also included. The discussions review topographic effects according to macro algae, micro algae and marine invertebrates. The inhibition or increased fouling by taxa is determined by comparing the settlement densities of textured surfaces with smooth controls where possible. Results of topographies that did not inhibit or encourage fouling are identified by the lack of statistical difference in settlement densities between tested topographies and smooth controls. Inconsistent settlement behaviour within a study or between studies is also noted.

1.2.1 Macro algae fouling

1.2.1.1 Micro scale

Ulva spores are quadriflagellated, pyriform microorganisms that are motile and selective about surfaces that are suitable for settlement [37–39]. Upon initial contact with the surface, a spore gauges potential for settlement by spinning in place while being anchored presumably by a temporary adhesive to the same spot. The spinning motion allows the spore to test its adhesion strength to the surface which depends on factors such as the surface's chemical properties. The longer the spinning duration, the more likely the spore will permanently attach to the surface [40, 41].

If the spore determines that the surface is suitable for permanent attachment, the spore moulds or conforms to the shape of the substratum and secretes a permanent adhesive to fix itself into position [42, 43]. Several studies mentioned that substrate depressions were good locations for spores that looked for refuge from hydrodynamic shear and grazing by predators [17, 32, 37, 44]. These studies also suggested that micro topography influenced the hydrodynamics of the surface by reducing fluid flow where depressions or crevices are present. Reduction of fluid flow decreases hydrodynamic stresses within these textures and increases the likelihood of retaining spore settlements.

Based on the attachment point theory presented in several studies, *Ulva* spores showed enhanced attachment when there were high levels of contact areas with topography features that consisted of depressions (i.e. between grooves or channels) [15, 24, 45, 46]. Substrates with uniform topographies that were 5 μm in width, length, height and distance apart generally encouraged algal spore settlement (Additional file 2: Table S2). *Ulva* spores have diameters of 5 μm and exhibited enhanced settlement densities in substrate depressions that were precisely 5 μm because surface area of contact between the substrate and spores were high [14, 15]. The support from surrounding topography features also reduced the amount of adhesive the spores would need to secrete to adhere within the substrate effectively (Figure 1A). 5 μm topographies also provided spores with a refuge from flow shear and did not effectively discourage spore settlements. When exposed to high hydrodynamic stresses from a water jet, only approximately half of the settled spores were removed [26].

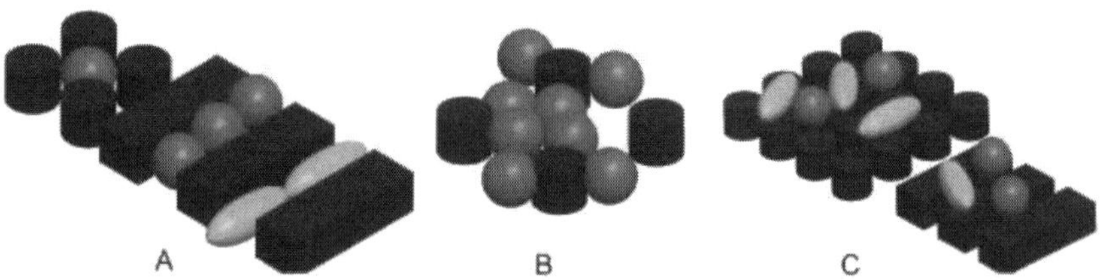

Figure 1. Settlement of organisms on topographically varied structures. **(A)** Organisms (green and light green) fitting well between adequately spaced topographical features. **(B)** Organisms leaning on each other and on topographies that are spaced too far apart. **(C)** 'Bridging' of organisms on multiple topography features.

Substrates with micro topography that were larger than 5 μm also encouraged *Ulva* spore settlement but were less effective in comparison to features that were precisely 5 μm [26, 47]. Under hydrodynamic stresses, the retention of spore settlements in features larger than 5 μm decreased as topography features increased in scale or distance apart. Spores also tended to 'lean' on nearby substrate features to increase its area of contact and reduce the effort required to adhere to the substrate [15]. This behaviour also explains the spores reduced preference to settle on smooth surfaces and the tendency of spores to lean on each other for support if there is no other choice but to settle (Figure 1B) [14]. This is demonstrated in investigations where topographies with average roughness heights (R_z) that ranged from 20-50 μm retained more spore settlements than similar topographies with roughness heights from 50-100 μm [17, 20, 33]. The larger exposed surface area of topographies with roughness heights from 50-100 μm resulted in a lower net surface area of contact between spores and the substrate.

Therefore, the amount of adhesives that spores had to secrete to adhere to the substrate effectively needed to increase.

Substrates with topography features that were less than 5 μm in dimensions were more effective at inhibiting spore attachment because spores were less able to conform into smaller sized depressions. As an effort to increase its contact area with these substrates, spores tended to be 'bridged' among multiple topographical features (Figure 1C). Spores could have landed and temporarily attached in attempts to conduct its initial assessment of the surface (i.e. spinning). However, it would be more likely that the complexity of the topography would have caused most spores to eventually leave and continue exploring [40]. Topography texture at this scale exhibited a level of 'tortuosity' which made the surface less attractive to spores for settlement. Topography tortuosity is a surface roughness parameter that describes the degree of freedom for organism movement on a textured surface (*df*). The more organism movement permitted over the topography, the more difficult it becomes for the organism to decide on an appropriate settlement space and the more 'tortuous' the substrate (Figure 2). It is highly likely that the spore would have rather released temporary adhesion and moved on to more suitable settlement sites. The degree of freedom of movement is one of three parameters associated with the dimensionless ratio derived by Schumacher et al. [16] known as the Engineered Roughness Index (ERI). Other variables that define this ratio include Wenzel's roughness factor (r), and the fraction of depression on the surface (f_D).

$$ERI_I = \frac{r \times df}{f_D} \tag{1}$$

Studies showed that spore settlement density decreased as the ratio increased [16, 29]. Micro-riblets with periodical lengths (4–12 μm), fixed widths of 2 μm and spaced at 2 μm apart had an ERI ratio of 9.5 and were the most effective at discouraging spore settlements among other topography features when compared with smooth control surfaces with an ERI ratio of 2. The ERI ratio provided a rough quantitative description of topography roughness and complexity but is limited because it is not inclusive of the height limit that is most effective at reducing spore settlement density.

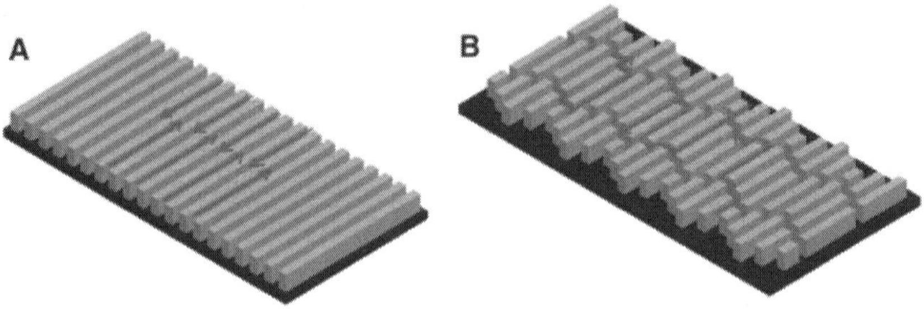

Figure 2. Tortuosity of topography with different complexity. Grooves in **(A)** are less 'tortuous' because organisms perceive the option to move only along one axis (indicated by red arrows); while riblets in **(B)** have higher 'tortuosity' because organisms are allowed the option to explore the surface along several directions.

Further investigation by Schumacher et al. [48] on topography of the same micro scale (less than 5 µm) determined that the aspect ratio (feature height/feature width) of topographical features influenced the density of spore settlements [14,48, 49]. The same micro riblets tested in Schumacher et al. [16] with the height of 3 µm exhibited the highest antifouling properties when compared to riblets that were 1 or 2 µm high. 3 µm high riblets significantly reduced contact between the spores and the floor of the substrate which reduced effective attachment of the spore to the surface (Figure 3).

A recent investigation of hexagonal pits that gradiently changed to zigzags (1- 10 µm) showed that *Ulva linza* spores preferentially settled at 'kink sites' that provided more contact area for attachment [43]. Settlement density of spores decreased as pit size decreased and when hexagonal pits changed into zigzags (Figure 4A). Apart from being larger than a spore's average size, when the hexagonal pits reached a width of 8.5 µm and changed to zigzags, the number of 'kink sites' reduced from 6 sites to only 2 sites (Figure 4A). 'Kink sites' offered more contact area for spores to mould against and increased their effective attachment (Figure 4B). The reduced number of 'kinks' on zigzags also reduced the number of preferential sites for spores to settle which reduced the overall density of spores settlements. Unlike uniform micro-topographies, the settlement density of spores increased as Wenzel's roughness (*r*) increased. This contradicts the ERI model (equation 1) which demonstrated that the settlement density of spores decreased with increasing Wenzel's roughness. Hence, the ERI model is not suitable to quantify the efficacy of topographies with highly irregular or gradiently changing dimensions.

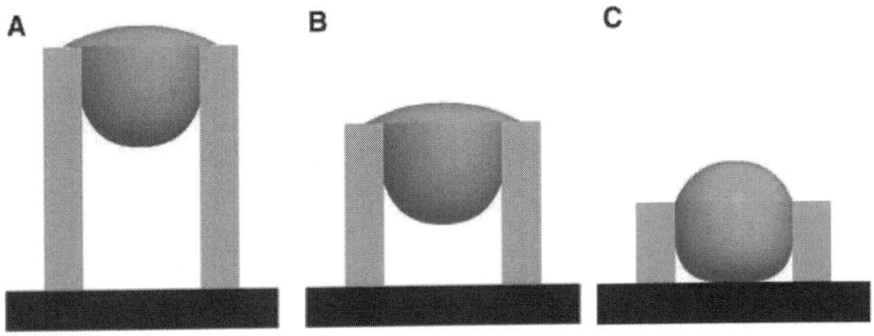

Figure 3. Effect of topography feature height towards effective spore attachment. Spore adhesion is the most effective in **(C)** because there is significant contact between spore and surrounding areas of the topography feature (i.e. walls, floor). Spore settlement in**(A)** and **(B)** is less effective since topography feature height it too high to allow spores more contact with its surroundings, especially the floor of the substrate [15].

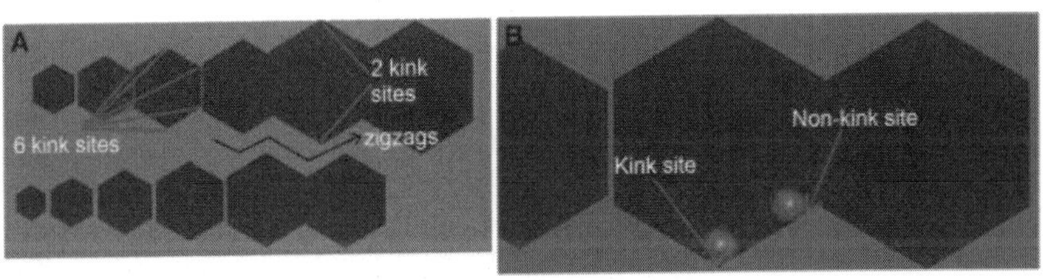

Figure 4. Kink sites for gradient hexagonal pits that change to zigzags. **(A)** As hexagonal pits change into zigzags, the number of 'kink sites' change from 6 sites to 2 sites when. **(B)** Spores preferred to settle in 'kink sites that provided a higher area of contact from surrounding walls and floors in comparison to 'non-kink sites'.

In contrast to *Ulva* spores, *Centroceras clavulatum* and *Polysiphonia sphaerocarpa* spores exhibited passive settlement behaviours on substrates but were less inclined to settle on textured substrates [24]. When compared to smooth controls, micro-sized sine wave grooves that ranged from 4 – 512 μm showed reduced densities of *C. clavulatum* spores and*Polysiphonia* spores. The lack of difference in *C. clavulatum* spore settlement densities on the same grooves of differing wavelengths suggests that the scale of micro topographical features did not affect the spores' settlement behaviour. Conversely, *P. sphaerocarpa* spores exhibited random trends of settlement density across grooves of differing wavelengths. Even though micro topography reduced the settlement densities of both spores, there is unfortunately insufficient understanding of their settling mechanisms to explain the difference in their settlement trends [50].

1.2.1.2 Hierarchical topographies

Hierarchical structures of micro peaks on irregular micro-cells reduced *Chlorella* (9-10 μm) and *Nannochloropsis maritima*(1-2 μm) settlements [51]. The micro peaks were 0.3 μm wide with an average height of 2 μm and were densely packed on cells that had an average width of 30 μm. This topography exhibited antifouling properties because gaps between peaks were too small (2-3 μm) to accommodate high settlement densities of both types of spores. 'Bridging' of spores between peaks also indicated that this topography was high in tortuosity which made it difficult for spores to establish contact with the substrate to adhere effectively. The difficulty to attach would have made the substrate less 'attractive' to the spores for settlement. Hence, results indicated that only sufficiently large gaps between micro cells that were not covered with micro peaks retained more spore settlements.

In contrast, hierarchical structures of periodical micro- and nano-sized sine-wave wrinkles that ranged from 5 nm to 500 μm enhanced *Ulva rigida* spore settlements [23]. Spores settled abundantly especially in depressions that were larger than the spore's body size and a majority of these features retained spore settlements even after being rinsed with water. Hierarchical structures of micro riblets with the same dimensions in the study by Schumacher et al. [16] layered on top of micro rectangular grooves (20 μm or 200 μm) also enhanced the same spores to settle [48]. Closer analysis of the laboratory assays under a Scanning Electron Microscope (SEM) determined that spaces between riblets and adjacent walls of grooves or areas of grooves that were not patterned with riblets were the areas that permitted spores to settle.

1.2.2 Microalgae fouling

1.2.2.1 Macro topographies

Field studies on microalgae communities of *Licmophora* sp. and other unidentified species of diatoms settled successfully on macro sized bumps that ranged from 1-5 mm (Additional file 2: Table S2). Settlement densities were particularly abundant on the highest elevations of these bumps [52]. Diatoms exhibit passive settlement onto a substrate and subsequent motility only when physical contact is made with the substrate to select preferred attachment sites [46]. Therefore, diatoms could have settled passively onto random areas on the substrate and migrated to elevated areas of bumps which would be the most strategic position for diatoms to reach sunlight and other nutrient supplies. Based on a study by Verran et al. [53], passive retention of micro-sized diatoms should be minimal on significantly larger macro-sized bumps. However, the presence of a biofilm layer could have conditioned the textured surface and provided a complex matrix that protected the diatoms from turbulent shear forces at the peaks of the bumps.

1.2.2.2 Micro topographies

The implications of micro topographies were evident when topographical feature sizes were proximate to the width of the diatoms' bodies [54]. Since tests on diatoms were limited to observing passive settlement on topographied surfaces with laboratory assays, hydrodynamic stresses would have not been consequential to the results. Although the ERI ratio has yet to be tested on diatoms, this ratio could be applicable to these organisms since both diatoms and spores are rather similar in size.

Based on the attachment point theory, *Nitzschia* cf. *paleacea* (2 μm) and *Fallacia carpentariae* (1 μm) settled on sine-wave grooves that were 4 and 2 μm in wavelength because multiple contact points were formed between the diatoms' bodies and the topographies. Cylindrically shaped diatoms were able to fit securely in the recessed areas of the grooves especially when the diatoms' bodies were aligned longitudinally to the grooves (Figure 1A) [46].

In contrast, the densities of larger diatoms *Amphora* sp. (7 μm) and *Navicula jeffreyi* (4 μm) were reduced on the same grooves because only an estimated maximum of six contact points were formed between these diatoms and the grooves. Both diatoms showed the tendency to be 'bridged' across several grooves and were not slim enough to fit securely in between the peaks of grooves.

The settlement densities of *Amphora* sp. were reduced on micro sized sine wave grooves when compared to smooth surfaces. This study also observed that sine wave grooves larger than diatoms showed slightly more diatom densities than grooves that were smaller than the diatoms' size, but these increases were considered statistically insignificant.

Apart from supporting the attachment point theory, it is possible that this observation also concurs with a derivation of the ERI ratio (equation 1) where the parameter that accounts for surface tortuosity is replaced with *n,* which is defined as the number of distinct features that is visible on the textured surface [29].

$$\mathrm{ERI}_{II} = \frac{r \times n}{f_D} \qquad (2)$$

From equation (2) substrates with constant surface areas would fit more grooves with smaller widths in comparison to larger width grooves. Similar to equation (1), the antifouling efficacy of the textured surface is proportional to the ERI value. Therefore, the slightly better antifouling performance of smaller width sine-wave grooves at resisting Amphora sp. settlement is warranted.

1.2.3 Invertebrate fouling

1.2.3.1 Macro topographies

Most macro invertebrates reviewed in Additional file 3: Table S3 settled effectively on macro-sized topographies that consisted of bumps, holes and V-shaped grooves [22, 36, 55–57]. These topographies were generally larger in scale relative to the size of invertebrates and were likely considered to be suitable by these organisms for settlement and adhesion. Turbulent flows over macro sized topographies creates disturbances in the viscous sublayer of the boundary layer that is directly adjacent to the surface [44]. Disturbance in the viscous sublayer increases hydrodynamic drag or friction on the surface of the substrate which is a condition that must be avoided particularly in the shipping industry. Hence, it could be argued that topographies at this scale are not suitable for marine antifouling applications. However, the differences in macro-invertebrate settlement patterns observed on macro-scale topographies highlight essential factors that are of interest to antifouling research.

The preference of *Crambe crambe* and *Scopalina lophyropoda* sponge larvae (average size of 1 mm^2) to settle in 5 mm wide V-shaped grooves large enough to shade them suggests that these larvae were affected by light. This is an example of an instance where negative phototaxis cues affected the distribution of sponge larvae [58]. It is also likely that barnacle cyprids, ascidians and bryozoan larvae tend to settle at the base of bumps or in large holes to take refuge from hydrodynamic shear (rheotaxis) and protection from desiccation, being grazed over or eaten by predators [35, 59]. Another probable reason for this observation is that hydrodynamic forces from turbulent water flow could have also caused these larvae to passively settle within these holes. However, holes with heights of 2 mm lacked *B. neritina* settlement because the average height of this bryozoan (2.8 mm) exceeded the height of these holes and significantly increased its exposure to predation.

Unlike *B. neritina*, the hydroid *Tubularia crocea* exhibited preference for the high areas of macro topographies that were most exposed to hydrodynamic shear stress and turbulence [60]. The reasons for this behaviour were uncertain but it is possible that the turbulent areas of macro topographies act as sediment traps that supply nutrients to the hydroid [52]. If the flow over the substrate is an ambient current, there is also the possibility that the water flow could have carried *T.crocea* larvae passively to the highest elevations of the topography instead of dragging larvae into protected crevices [61]. Macro holes seemed to be effective at preventing *T. crocea* settlements but the reasons for this finding were not substantiated. Among the possibilities for this observation includes rejection of holes by the hydroid larvae or that macro holes might have been too small to fit adult hydroids. It is also likely that the formation of eddies in the holes from turbulent fluid flow could have prevented larvae from settling successfully.

The bryozoan *Membranipora membranacea* has a flexible body that is flat which often puts it at risk of being encrusted over by competing organisms that settle in the same vicinity [36]. This encourages the bryozoan to respond to spatial refuge cues and seek the highest elevations of topographical features [36]. This explained their preference to settle on the top of macro bumps and on the outer rims of macro holes where the probability of getting encrusted over is lower.

Some species showed gregarious settlement behaviour responding to conspecific cues [34]. Barnacle cyprids and pediveligers of mussels respond positively to conspecific cues left by preceding or current colonies of their respective species. These larvae have 'feet’ that explore substrates and settle when they detect familiar chemical compounds, the physical presence of living or deceased predecessors and in certain cases, even chemical secretions from their prey.

Many studies have shown that biofilms provide an important settlement cue for macro invertebrates. Biofilms layers consist of polymeric substances, microorganisms, and miscellaneous matter that can attract many invertebrates to settle. For example, larvae of *Hydroides elegans*, and *Mytilus galloprovincialis* exhibit positive settlement on substrates that were coated with layers of biofilms that contain chemical substances that are familiar to them [62–64]. Biofilm layers that are formed under different hydrodynamic conditions also affected the attachment strength of cyprids. Biofilms formed under high shear stresses are more densely linked than biofilms formed under low shear stresses. The denser matrix of biofilms that grew in high shear conditions allowed cyprids to attach more firmly in comparison to substrates with biofilms that grew under low shear stresses [65, 66].

1.2.3.2 Micro topographies

Like *Ulva* spores, there are studies that explained the settlement mechanisms of barnacles [40, 67]. A barnacle nauplius matures into cyprids that swim about in the water column looking for a place to settle. When it comes into contact with a substrate, it tracks or walks about leaving 'footprints’ to assess the suitability of the surface for settlement. While it has been mentioned that cyprids respond to conspecific cues, surface topography also contributes to the settlement trends of these organisms [22]. Macro topographies are favourable for cyprids to settle but comparisons between studies have shown that significantly smaller micro topographies were effective at deterring cyprids from settling [22, 68]. Many studies have tested the effects of micro topographies on the settlement behaviour of barnacle cyprids and these studies consolidate the importance of topography scale in relation to the size of the cyprids [48, 69–71].

Regardless of geometry, micro topographies with geometry widths that ranged from 1-5 μm and approximately 64-256 μm reduced cyprid settlements (Additional file 3: Table S3). Cyprids track or walk about with 'feet' before settling permanently and the tips of their 'feet' consist of ovoid antennular discs that are approximately 20-30 μm [70]. It could be argued that topographies ranging from 1-5 μm were rejected by cyprids because there were insufficient contact points to support settlement and attachment. Likewise, topographies that ranged from 64-256 μm would have been too small to accommodate metamorphosed juvenile barnacles with body widths of approximately 500 μm.

Micro topographies that ranged from 8-64 μm and 500-1000 μm enhanced cyprid settlements. Sine wave grooves that were 16 μm in wavelength were very supportive of *B. amphitrite* attachment which suggests that this size could be a critical dimension for this particular cyprid species. However, it is difficult to ascertain the reasons for this finding and further studies are warranted to determine this. It is probable that similar grooves which were 32-64 μm provided strong anchor points for attachment because cyprid antennular discs (30 μm) were compatible in size to these grooves. Topographies that were 500-1000 μm were expected to encourage cyprids to settle because cyprids were close in size and would be able to attach easily onto topographies of this range [70].

A study that compared between V-shaped grooves and peaks of the same structure height (69 μm) and base size (97 μm) with smooth controls found that both micro topographies reduced *Balanus improvisus* cyprid settlement. However, based on the overall settlement density of cyprids, grooves were more effective than peaks at resisting cyprid settlements [32]. A separate study found that micro-sized sine wave grooves with peak wavelengths of 1.5-2 μm were better than roughened grooves of similar geometry and magnitude at deterring *Semibalanus balanoides* cyprids from settling [68]. The reduction in settlement density of cyprids in both studies can be attributed to topography size being too small to accommodate the larger bodies of cyprids (500-1000 μm). However, the higher antifouling performance of V-shaped grooves and unroughened sine-wave grooves in both studies respectively also suggests that cyprids are likely discouraged to settle on micro textures that have a degree of symmetry and isotropicity (Figure 5). These results also explain the enhanced antifouling efficacy of isotropic micro mesh patterns at preventing cyprids from settling in a number of investigations [33,71, 72].

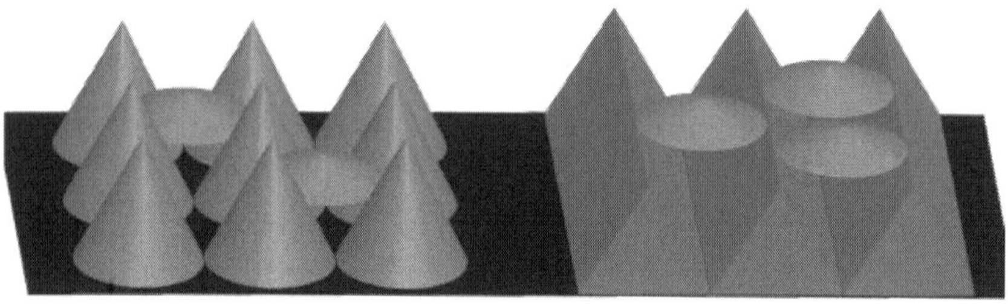

Figure 5. Influence of isotropy on similarly scaled but geometrically different topography. The illustration shows that peaks are better than the more isotropically inclined V-shaped grooves of approximate size at providing secure settlement points for the attachment of cyprid base plates (beige).

Although there are fewer studies of micro topographies on other marine invertebrates, these studies also indicated that micro topographies have varying effects on other larvae settlements. Larvae of *Bugula neritina* have diameters that are approximately 300-400 μm which explains the reduced settlements on sine wave grooves that were only 256 μm wide. As expected, the largest sine wave grooves (512 μm) were able to fit the zooids of this bryozoa. The increased settlement densities of *B. neritina* larvae on smaller grooves that ranged from 8–64 μm were unexpected and difficult to explain. Grooves of this range have an overall exposed surface area that is greater than larger sine wave grooves. Therefore, it is possible that the amount of adhesives adsorbed on grooves of this roughness range would be high and could have aided in retaining more bryozoan larvae [44].

Laboratory assays found that rectangular grooves ranging from 100-1000 μm enhanced the passive settlement of *Mytilus galloprovincialis* pediveligers when compared to smooth controls [5]. The settlement trend of pediveligers indicated sensitivity to texture size of micro topographies. The average body size of *Mytilus galloprovincialis* pediveligers are approximately 190 μm and 260 μm in width and length respectively [73]. Unlike *Ulva* spores, pediveligers are fixed in shape and would find it difficult to conform or fit into grooves that are near to its size. This substantiates the reduction in settlement on grooves that ranged from 100-250 μm.

The single flexible 'foot' of pediveligers is approximately 40 μm in diameter which is small enough to fit square grooves that were 40-80 μm in width and height [25, 73]. These studies also determined that pediveligers showed poor retention on topographies of this range when subjected to hydrodynamic shear tests. One possibility for this observation is that settled pediveligers could have metamorphosed to sizes that could no longer fit these grooves within a matter of hours after initial settlement and be removed easily when rinsed with water. Several separate studies also observed that topographies with heights that ranged from 40-100 μm showed a reduction in retained settlements of barnacle cyprids, *Mytilus edulis* larvae and *Ulva* spores after being subjected to hydrodynamic shear [17, 25, 32, 48]. It can be observed that topography with roughness heights within this range are effective at reducing invertebrate larvae and algae spore settlements if the textured surface is subjected to hydrodynamic shear. It is likely that the resultant drag and lift subjected to these topographies reduced the adhesion strength of these fouling organisms. From these studies, it is possible to postulate that the maximum cutoff height that would be most effective at discouraging fouling by the aforementioned larvae could be approximately 100 μm.

The encrusting tube worms *Hydroides elegans* and *Polydora* sp. have bodies that are approximately 100-250 μm wide and also exhibited settlement behaviours that concurred with the attachment point theory [24, 45]. Scardino et al. [24] observed increased *H. elegans* larvae settlement on sine wave grooves with wavelengths of 125 μm and above while Bers and Wahl [25] found that sine wave grooves less than 5 μm reduced settlements of the same larvae. Bers and Wahl [25] also observed reduced densities of *Polydora* sp. larvae on rectangular grooves with widths that ranged from 30 – 50 μm.

Several field studies found inconsistent settlement behaviours of *Polydora* sp. larvae on micro bumps that range from 10-500 μm [25, 52]. Reasons for this behaviour were not clear but environmental factors such as the presence of biofilms, conspecific cues and hydrodynamic stresses on the textured surface could have contributed to this inconsistency. This result also indicates that environmental cues in natural settings might have more influence than micro topographies towards polychaete larvae settlement behaviour.

Both these studies were the only studies that observed the effects of topography towards ciliates which makes it difficult to ascertain the reasons for their settlement behaviour. Nonetheless, with the exception of *Vorticella* sp., it is likely that other ciliates were less inclined to settle on textured or roughened substrates across the macro, and micro scales (Additional file3: Table S3).

1.2.3.3 Hierarchical topographies

Hierarchical structures that consisted of periodical sine wave wrinkles (5 nm-500 μm in wavelength) effectively deterred cyprids while micro bumps (200 μm in diameter) overlapped with smaller micro peaks (1-5 μm in width and height) reduced cyprid and mussel settlement [23, 25]. In these cases, the topographies were often too small to fit both types of larvae. Since both investigations were field studies, it is also likely that hydrodynamic stresses on this hierarchical surface would have contributed to the antifouling efficacy of these topographies.

Another possibility for the enhanced antifouling performance of hierarchical topographies influenced the surface exploration time of cyprids and pediveligers which discouraged these larvae to settle and attach. Several studies have shown that reducing the exploration time of cyprids and mussel larvae decreased attachment and also increased the rejection rate of the substrate for attachment and metamorphosis [20, 33, 72].

1.2.4 Nano topography

Antifouling research applying nano-sized topographies is recent and technically more complicated. The AMBIO collaboration was an effort that investigated the potential of nanostructures for antifouling applications [74]. These studies involved the development of nanostructures with novel, non-toxic materials to develop better antifouling solutions. Apart from concepts of superhydrophobicity, surface energy and surface wettability, there are yet to be established theories that can confirm the direct correlation of the antifouling properties of nano-topographies with the settlement trends of marine fouling organisms [74–76]. Nonetheless, with the exception of marine bacteria, settlement densities of other taxa were reduced on substrates with topographies of this scale (Additional file 2: Table S2, Additional file 3: Table S3).

There are not many studies that discussed the effect of topography on the settlement behaviour of marine bacteria to determine a distinct trend that could explain their behaviour during settlement and attachment [77–79]. Several studies that analysed the settlement behaviour of marine bacteria on substrates did not indicate that topography had a significant effect on their settlement behaviour. Nonetheless, marine bacteria density increased on substrates with increased nano-scale surface roughness particularly where the measured root mean square roughness (R_q) equals to 0.01 μm [78, 80]. This is likely because nano-rough surfaces have increased surface area for adhesion of extracellular polymeric substances (EPS) which improved the attachment of marine bacteria [45].

The greater reduction of diatoms and spores on substrates of nano peaks compared to nano bumps suggests that sharper nano peaks might have provided less support than broad nano bumps for attachment [81]. Similar to observed settlement trends of marine bacteria, several studies proposed that nanostructured substrates influenced surface wettability which influences organism adhesion. For example, the hydrophobic nature of adhesives secreted by diatoms makes them less inclined to adhere to hydrophilic nanostructures [6, 82]. In contrast, *Ulva* spores attached weakly on hydrophobic nanostructures due to the hydrophilic nature of its adhesive [42].

Another proposed antifouling mechanism involving nano-topographies involved the surface reconstruction of nanostructures after immersion in water [6, 83–86]. This shift in surface morphology might contribute to varying surface wettability of the substrate and influence organism adhesion. There has yet to be sufficient evidence or studies to support or deny these theories. The exact antifouling mechanism of nano-topographies remains unfounded and still lacks adequate understanding.

Several studies have argued that nano-topographies created trapped 'air pockets' between the solid–liquid interface that impedes larvae or microorganisms from having direct contact with the substrate [38, 80, 87–89]. The nature of these substrates was described as superhydrophobic and commonly characterised by measured contact angles that are at least 135° [90]. Due to their significantly larger bodies, it is unlikely that macro invertebrate larvae would be able to sense undulating irregularities at the nano-scale. Topography with features that are this small would not be large enough to induce significant drag or skin friction beyond the viscous sublayer of the water column in direct contact with the substrate. It is more probable that the 'air pockets' reduced macro invertebrate larvae density by preventing them from having physical contact with the substrate.

This finding is consistent with a separate field study which found that superhydrophobic nano-substrates prevented macro and microorganisms from settling for a period of 6 months [91]. In many cases, there has yet to be a solution which allows these substrates to maintain their impervious nature for prolonged periods of time. Subsequent fouling occurred because 'air pockets' eventually dissolved in water and also because the substrate was damaged by grazing from marine organisms.

1.2.5 Future directions

The critical range for microorganisms and macro organisms to settle differ significantly. This increases the difficulty to identify a range that would be effective at preventing both micro and macro organisms from settling. Micro topographies that usually consist of one uniform arrangement of geometry and scale have shown limited success at stopping settlement and did not have the same level of efficacy at inhibiting a wider range of taxa.

Hierarchical topographies are relatively new and have yet to be extensively explored for marine antifouling applications. Nonetheless, there were promising results which indicated that these topographies might have the potential to reduce fouling by both micro and macro organisms. Therefore, future studies on antifouling topographies should consider mixed scales of hierarchical structures that might have the potential to prevent a wider array of taxa from settling.

Unlike macro and micro topographies, nanostructures that were tested for marine antifouling purposes have been limited to irregular self-assembled geometries of mixed nano sizes. There are studies that have studied the effects of regular arrays of nano topographies at deterring biofouling but these have been limited to other applications [92, 93]. Hence, future investigations should develop regular, nano-sized topographies to be analysed for marine antifouling purposes.

Surface topography also affects hydrodynamic stresses exerted on their ambient surroundings which also influence the settlement of fouling organisms. It would be beneficial to conduct more field assays taking this and other environmental factors into consideration especially on less explored nano or hierarchical topographies. Laboratory assays and field trials often yield results that differ from each other. This indicates the influence of environmental variables such as biofilm growth and conspecific cues on organism settlement trends [25, 94].

A recent study by Vucko et al. [94] determined that the settlement trends of a range of marine organisms differed when comparing results from laboratory assays and field trials. It is speculated that biofilm growth could have contributed to the differences in observations. Biofilm formation on submerged substrates could have rendered the antifouling properties of topographies irrelevant. The next best course of action would be to combine antifouling topography with other antifouling mechanisms such as surface energy to devise better solutions and reduce biofilm formation [95]. It is probable that only combinations of several mechanisms could result in more robust topographies with better antifouling abilities.

This might now be possible with improvements made to the ERI_{II} ratio (equation 2) to quantify the combined effects of substrate material (i.e. surface energy) and topography to optimise future designs of antifouling solutions [29–31]. Decker et al. [96] further improved these results to predict not just the inhibitory but also the enhancing aspects of foul resistant topographies that influence organism settlement trends. These quantitative models are necessary and useful to evaluate the performance of foul-resistant topographies and also to validate results from laboratory and field trials. Future studies on these models could consider a wider range of environmental variables such as salinity and temperature to make the results more comparable to experimental trials.

2 CONCLUSIONS

Macro topographies encouraged most organisms to settle and are not suitable to be applied as antifouling topographies. In general, sub-macro topographies exhibited antifouling properties and there were encouraging results but there has yet to be a topography configuration capable of reducing the settlement of a wide range of marine taxa. Closer inspection of micro topographies determined that the topography scale in terms of height/depth, width, symmetry, distance apart from each other and average roughness were important parameters that can have polarising effects on different taxa.

As it stands, it is difficult and possibly not practical to develop optimum topographies with scales that can deter the settlement of all types of taxa. Nonetheless, uniformly patterned nano-topographies and hierarchical topographies have yet to be widely explored and might have the potential to deter fouling from a wider range of marine taxa and should be investigated further. Both scales of topography also warrant further investigations to clarify their exact antifouling mechanisms to increase the antifouling efficacy of both types of topography.

A solution is also required to increase the durability of antifouling topographies in marine environments to be considered a viable option for commercial antifouling applications. Moreover, there has yet to be any solution to prevent the formation of biofilms that would eventually hide and overcome the antifouling properties of topographies. This suggests that environmental factors (e.g. hydrodynamic shear, biofilms, conspecific cues) are influential towards antifouling topographies and current topographies are still not well equipped to be in marine environments for long periods of time. Hence, combinations of several strategies are likely the best option to maintain the durability and performance of antifouling topographies. More field trials are necessary to corroborate observations from laboratory tests and calculations from quantitative models for optimum designs of foul-resistant topographies.

3 AUTHORS' INFORMATION

Co-authors: James Walker and Odette Paramor.

COMPETING INTERESTS

The authors declare that they have no competing interests.

AUTHORS' CONTRIBUTIONS

FWYM drafted the manuscript, figures and tables under the frequent advisement of OP and JW. Both OP and JW proof read the manuscript and all authors agree with the final version of the manuscript.

REFERENCES

1. Schultz MP, Bendick JA, Holm ER, Hertel WM: **Economic impact of biofouling on a naval surface ship.** Biofouling 2010, **27:**87–98.

2. Fitridge I, Dempster T, Guenther J, de Nys R: **The impact and control of biofouling in marine aquaculture: a review.** Biofouling 2012,**28:**649–669.

3. Strand J, Jacobsen JA: **Accumulation and trophic transfer of organotins in a marine food web from the Danish coastal waters.** Sci Total Environ 2005, **350:**72–85.

4. Evans SM, Leksono T, McKinnell PD: **Tributyltin pollution: a diminishing problem following legislation limiting the use of TBT-based anti-fouling paints.** Mar Pollut Bull 1995, **30:**14–21.

5. Carl C, Poole AJ, Vucko MJ, Williams MR, Whalan S, de Nys R: **Enhancing the efficacy of fouling-release coatings against fouling by Mytilus galloprovincialis using nanofillers.** Biofouling 2012, **28:**1077–1091.

6. Fang J, Kelarakis A, Wang D, Giannelis EP, Finlay JA, Callow ME, Callow JA: **Fouling release nanostructured coatings based on PDMS-polyurea segmented copolymers.** Polymer 2010, **51:**2636–2642.

7. Xie L, Hong F, He C, Ma C, Liu J, Zhang G, Wu C: **Coatings with a self-generating hydrogel surface for antifouling.** Polymer 2011,**52:**3738–3744.

8. Ekblad T, Bergström G, Ederth T, Conlan SL, Mutton R, Clare AS, Wang S, Liu Y, Zhao Q, D'Souza F, Donnelly GT, Willemsen PR, Pettitt ME, Callow ME, Callow JA, Liedberg B: **Poly(ethylene glycol)-containing hydrogel surfaces for antifouling applications in marine and freshwater environments.** Biomacromolecules 2008, **9:**2775–2783.

9. Akesso L, Navabpour P, Teer D, Pettitt ME, Callow ME, Liu C, Su X, Wang S, Zhao Q, Donik C, Kocijan A, Jenko M, Callow JA: **Deposition parameters to improve the fouling-release properties of thin siloxane coatings prepared by PACVD.** Appl Surf Sci 2009, **255:**6508–6514.

10. Lejars M, Margaillan A, Bressy C: **Fouling release coatings: a nontoxic alternative to Biocidal antifouling coatings.** Chem Rev 2012,**112:**4347–4390.

11. Magin CM, Cooper SP, Brennan AB: **Non-toxic antifouling strategies.** Mater Today 2010, **13:**36–44.

12. Grozea CM, Walker GC: **Approaches in designing non-toxic polymer surfaces to deter marine biofouling.** Soft Matter 2009, **5:**4088–4100.

13. Krishnan S, Weinman CJ, Ober CK: **Advances in polymers for anti-biofouling surfaces.** J Mater Chem 2008, **18:**3405–3413.

14. Carman ML, Estes TG, Feinberg AW, Schumacher JF, Wilkerson W, Wilson LH, Callow ME, Callow JA, Brennan AB: **Engineered antifouling microtopographies – correlating wettability with cell attachment.** Biofouling 2006, **22:**11–21.

15. Callow ME, Jennings AR, Brennan AB, Seegert CE, Gibson A, Wilson L, Feinberg A, Baney R, Callow JA: **Microtopographic cues for settlement of zoospores of the green fouling alga Enteromorpha.** Biofouling 2002, **18:**229–236.

16. Schumacher JF, Carman ML, Estes TG, Feinberg AW, Wilson LH, Callow ME, Callow JA, Finlay JA, Brennan AB: **Engineered antifouling microtopographies – effect of feature size, geometry, and**

roughness on settlement of zoospores of the green alga Ulva. Biofouling2007, **23:**55–62.

17. Granhag LM, Finlay JA, Jonsson PR, Callow JA, Callow ME: **Roughness-dependent removal of settled spores of the green alga ulva (syn. Enteromorpha) exposed to hydrodynamic forces from a water Jet.** Biofouling 2004, **20:**117–122.

18. Baum CB, Meyer WM, Stelzer RS, Fleischer LGF, Siebers DS: **Average nanorough skin surface of the pilot whale (Globicephala melas, Delphinidae): considerations on the self-cleaning abilities based on nanoroughness.** Mar Biol 2002, **140:**653–657.

19. Salta M, Wharton JA, Stoodley P, Dennington SP, Goodes LR, Werwinski S, Mart U, Wood RJK, Stokes KR: **Designing biomimetic antifouling surfaces.** Philos T Roy Soc A 2010, **368:**4729–4754.

20. Berntsson KM, Jonsson PR, Lejhall M, Gatenholm P: **Analysis of behavioural rejection of micro-textured surfaces and implications for recruitment by the barnacle Balanus improvisus.** J Exp Mar Biol Ecol 2000, **251:**59–83.

21. Chun Chaw K, Dickinson GH, Yang Ang K, Deng J: **Surface exploration of Amphibalanus amphitrite cyprids on microtextured surfaces.** Biofouling 2011, **27:**413.

22. Hills JM, Thomason JC: **The effect of scales of surface roughness on the settlement of barnacle (Semibalanus balanoides) cyprids.** Biofouling 1998, **12:**57–69.

23. Efimenko K, Finlay J, Callow ME, Callow JA, Genzer J: **Development and testing of hierarchically wrinkled coatings for marine antifouling.** ACS Appl Mater Inter 2009, **1:**1031–1040.

24. Scardino AJ, Guenther J, de Nys R: **Attachment point theory revisited: the fouling response to a microtextured matrix.** Biofouling2007, **24:**45–53.

25. Bers AV, Wahl M: **The influence of natural surface microtopographies on fouling.** Biofouling 2004, **20:**43–51.

26. Hoipkemeier-Wilson L, Schumacher JF, Carman ML, Gibson AL, Feinberg AW, Callow ME, Finlay JA, Callow JA, Brennan AB: **Antifouling potential of lubricious, micro-engineered, PDMS Elastomers against zoospores of the green fouling alga ulva (Enteromorpha).**Biofouling 2004, **20:**53–63.

27. Thomason JC, Davenport J, Rogerson A: **Antifouling performance of the embryo and eggcase of the dogfish Scyliorhinus canicula.** J Mar Biol Assoc UK 1994, **74:**823–836.

28. Howell D, Behrends B: **A review of surface roughness in antifouling coatings illustrating the importance of cutoff length.** Biofouling 2006, **22:**401–410.

29. Long CJ, Schumacher JF, Robinson PAC, Finlay JA, Callow ME, Callow JA, Brennan AB: **A model that predicts the attachment behavior of Ulva linza zoospores on surface topography.** Biofouling 2010, **26:**411–419.

30. Magin CM, Long CJ, Cooper SP, Ista LK, López GP, Brennan AB: **Engineered antifouling microtopographies: the role of Reynolds number in a model that predicts attachment of zoospores of Ulva and cells of Cobetia marina.** Biofouling 2010, **26:**719–727.

31. Magin CM, Finlay JA, Clay G, Callow ME, Callow JA, Brennan AB: **Antifouling performance of cross-linked hydrogels: refinement of an attachment model.** Biomacromolecules 2011, **12:**915–922.

32. Petronis Š, Berntsson K, Gold J, Gatenholm P: **Design and microstructuring of PDMS surfaces for improved marine biofouling resistance.** J Biomat Sci-Polym E 2000, **11:**1051–1072.

33. Berntsson KM, Andreasson H, Jonsson PR, Larsson L, Ring K, Petronis S, Gatenholm P: **Reduction of barnacle recruitment on micro–textured surfaces: analysis of effective topographic characteristics and evaluation of skin friction.** Biofouling 2000, **16:**245–261.

34. Chabot R, Bourget E: **Influence of substratum heterogeneity and settled barnacle density on the settlement of cypris larvae.** Mar Biol 1988, **97:**45–56.

35. Walters LJ, Wethey DS: **Settlement and early post-settlement survival of sessile marine invertebrates on topographically complex surfaces: The importance of refuge dimensions and adult morphology.** Mar Ecol Prog Ser 1996, **137:**161–171.

36. Walters LJ, Wethey DS: **Settlement, refuges, and adult body form in colonial marine invertebrates: a field experiment.** Biol Bull 1991,**180:**112–118.

37. Figueiredo MAO, Norton TA, Kain JM: **Settlement and survival of epiphytes on two intertidal crustose coralline alga.** J Exp Mar Biol Ecol 1997, **213:**247–260.

38. Cao X, Pettitt ME, Wode F, Arpa Sancet MP, Fu J, Ji J, Callow ME, Callow JA, Rosenhahn A, Grunze M: **Interaction of zoospores of the green alga ulva with bioinspired micro- and nanostructured surfaces prepared by polyelectrolyte layer-by-layer self-assembly.** Adv Funct Mater 2010, **20:**1984–1993.

39. Rosenhahn A, Finlay JA, Pettit ME, Ward A, Wirges W, Gerhard R, Callow ME, Grunze M, Callow JA: **Zeta potential of motile spores of the green alga Ulva linza and the influence of electrostatic interactions on spore settlement and adhesion strength.** Biointerphases2009, **4:**7–11.

40. Rosenhahn A, Sendra GH: **Surface sensing and settlement strategies of marine biofouling organisms.** Biointerphases 2012, **7:**1–13.

41. Heydt M, Pettitt ME, Cao X, Callow ME, Callow JA, Grunze M, Rosenhahn A: **Settlement behavior of zoospores of ulva linza during surface selection studied by digital holographic microscopy.** Biointerphases 2012, **7:**1–7.

42. Callow J, Callow M: **The ulva spore adhesive system.** In Biological adhesives. Edited by: Smith AM, Callow JA. Heidelberg: Springer Berlin; 2006.

43. Xiao L, Thompson SEM, Röhrig M, Callow ME, Callow JA, Grunze M, Rosenhahn A: **Hot embossed microtopographic gradients reveal morphological cues that guide the settlement of zoospores.** Langmuir 2012, **29:**1093–1099.

44. Koehl MRA: **Mini review: hydrodynamics of larval settlement into fouling communities.** Biofouling 2007, **23:**357–368.

45. Verran J, Boyd RD: **The relationship between substratum surface roughness and microbiological and organic soiling: a review.** Biofouling 2001, **17:**59–71.

46. Scardino AJ, Harvey E, De Nys R: **Testing attachment point theory: diatom attachment on microtextured polyimide biomimics.** Biofouling 2006, **22:**55–60.

47. Finlay JA, Krishnan S, Callow ME, Callow JA, Dong R, Asgill N, Wong K, Kramer EJ, Ober CK: **Settlement of ulva zoospores on patterned fluorinated and PEGylated monolayer surfaces.** Langmuir 2007, **24:**503–510.

48. Schumacher JF, Aldred N, Callow ME, Finlay JA, Callow JA, Clare AS, Brennan AB: **Species-specific engineered antifouling topographies: correlations between the settlement of algal zoospores and barnacle cyprids.** Biofouling 2007, **23:**307–317.

49. Schumacher JF, Long CJ, Callow ME, Finlay JA, Callow JA, Brennan AB: **Engineered nanoforce gradients for inhibition of settlement (attachment) of swimming algal spores.** Langmuir 2008, **24:**4931–4937.

50. Scardino AJ: Biomimetic fouling control PhD thesis. James Cook University; 2006. http://eprints.jcu.edu.au/2091/

51. Wan F, Pei X, Yu B, Ye Q, Zhou F, Xue Q: **Grafting polymer brushes on biomimetic structural surfaces for anti-algae fouling and foul release.** ACS Appl Mater Inter 2012, **4:**4557–4565.

52. Köhler J, Hansen PD, Wahl M: **Colonization patterns at the substratum–water interface: how does surface microtopography influence recruitment patterns of sessile organisms?** Biofouling 1999, **14:**237–248.

53. Verran J, Lees G, Shakespeare AP: **The effect of surface roughness on the adhesion of Candida albicans to acrylic.** Biofouling 1991,**3:**183–191.

54. Whitehead KA, Verran J: **The effect of surface topography on the retention of microorganisms.** Food Bioprod Process 2006, **84:**253–259.

55. Skinner LF, Coutinho R: **Effect of microhabitat distribution and substrate roughness on barnacle Tetraclita stalactifera (Lamarck, 1818) settlement.** Braz Arch Biol Techn 2005, **48:**109–113.

56. Gribben PE, Jeffs AG, de Nys R, Steinberg PD: **Relative importance of natural cues and substrate morphology for settlement of the New Zealand Greenshell™ mussel, Perna canaliculus.** Aquaculture 2011, **319:**240–246.

57. Bourget E, DeGuise J, Daigle G: **Scales of substratum heterogeneity, structural complexity, and the early establishment of a marine epibenthic community.** J Exp Mar Biol Ecol 1994, **181:**31–51.

58. Maldonado M, Uriz MJ: **Microrefuge exploitation by subtidal encrusting sponges: patterns of settlement and post-settlement survival.** Mar Ecol Prog Ser 1998, **174:**141–150.

59. Walters LJ: **Post-settlement success of the arborescent bryozoan Bugula neritina (L.): the importance of structural complexity.** J Exp Mar Biol Ecol 1992, **164:**55–71.

60. Lemire M, Bourget E: **Substratum heterogeneity and complexity influence micro-habitat selection of Balanus sp. and Tubulariacrocea larvae.** Mar Ecol Prog Ser 1996, **135:**77–87.

61. Abelson A, Denny M: **Settlement of marine organisms in flow.** Annu Rev Ecol Syst 1997, **28:**317–339.

62. Huang L, Hadfield : **Composition and density of bacterial biofilms determine larval settlement of the polychaete Hydroides elegans.** Mar Ecol Prog Ser 2003, **260:**161–172.

63. Hung O, Lee O, Thiyagarajan V, He H, Xu Y, Chung H, Qiu J, Qian P: **Characterization of cues from natural multi-species biofilms that induce larval attachment of the polychaete Hydroides elegans.** Aquat Biol 2009, **4:**253–262.

64. Satuito C, Shimizu K, Fusetani N: **Studies on the factors influencing larval settlement in Balanus amphitrite and Mytilus galloprovincialis.** Hydrobiologia 1997, **358:**275–280.

65. Neal AL, Yule AB: **The tenacity of Elminius modestus and Balanus perforatus cyprids to bacterial films grown under different shear regimes.** J Mar Biol Assoc UK 1994, **74:**251–257.

66. Neal AL, Simões FN, Yule AB: **Interactions between shear rates and biofilms affecting exploratory behaviour by cyprids ofElminius modestus (Cirripedia).** Mar Biol 1996, **127:**241–246.

67. Khandeparker L, Anil AC: **Underwater adhesion: the barnacle way.** Int J Adhesion Adhesives 2007, **27:**165–172.

68. Bers AV, Prendergast GS, Zürn CM, Hansson L, Head RM, Thomason JC: **A comparative study of the anti-settlement properties of mytilid shells.** Biol Lett 2006, **2:**88–91.

69. Olivier F, Tremblay R, Bourget E, Rittschof D: **Barnacle settlement: field experiments on the influence of larval supply, tidal level, biofilm quality and age on Balanus amphitrite cyprids.** Mar Ecol Prog Ser 2000, **199:**185–204.

70. Aldred N, Scardino A, Cavaco A, de Nys R, Clare AS: **Attachment strength is a key factor in the selection of surfaces by barnacle cyprids (Balanus amphitrite) during settlement.** Biofouling 2010, **26:**287–299.

71. Andersson M, Berntsson K, Jonsson P, Gatenholm P: **Microtextured surfaces: towards macrofouling resistant coatings.** Biofouling 1999, **14:**167–178.

72. Berntsson KM, Jonsson PR, Larsson AL, Holdt S: **Rejection of unsuitable substrata as a potential driver of aggregated settlement in the barnacle Balanus improvisus.** Mar Ecol Prog Ser 2004, **275:**199–210.

73. Carl C, Poole AJ, Sexton BA, Glenn FL, Vucko MJ, Williams MR, Whalan S, de Nys R: **Enhancing the settlement and attachment strength of pediveligers of Mytilus galloprovincialis bychanging surface wettability and microtopography.** Biofouling 2012, **28:**175–186.

74. Rosenhahna A, Ederth T, Pettitt M: **Advanced nanostructures for the control of biofouling: the FP6 EU integrated project AMBIO.**Biointerphases 2008, **3:**IR1-IR5.

75. Beigbeder A, Degee P, Conlan SL, Mutton RJ, Clare AS, Pettitt ME, Callow ME, Callow JA, Dubois P: **Preparation and characterisation of silicone-based coatings filled with carbon nanotubes and natural sepiolite and their application as marine fouling-release coatings.** Biofouling 2008, **24:**291–302.

76. Gunari N, Brewer LH, Bennett SM, Sokolova A, Kraut ND, Finlay JA, Meyer AE, Walker GC, Wendt DE, Callow ME: **The control of marine biofouling on xerogel surfaces with nanometer-scale topography.** Biofouling 2011, **27:**137–149.

77. Wouters M, Rentrop C, Willemsen P: **Surface structuring and coating performance: novel biocidefree nanocomposite coatings with anti-fouling and fouling-release properties.** Prog Org Coat 2010, **68:**4–11.

78. Mitik-Dineva N, Wang J, Truong VK, Stoddart PR, Malherbe F, Crawford RJ, Ivanova EP: **Differences in colonisation of five marine bacteria on two types of glass surfaces.** Biofouling 2009, **25:**621–631.

79. Navabpour P, Teer D, Su X, Liu C, Wang S, Zhao Q, Donik C, Kocijan A, Jenko M: **Optimisation of the properties of siloxane coatings as anti-biofouling coatings: Comparison of PACVD and hybrid PACVD–PVD coatings.** Surf Coat Technol 2010, **204:**3188–3195.

80. Scardino AJ, Zhang H, Cookson DJ, Lamb RN, Rd N: **The role of nano-roughness in antifouling.** Biofouling 2009, **25:**757–767.

81. Akesso L, Pettitt ME, Callow JA, Callow ME, Stallard J, Teer D, Liu C, Wang S, Zhao Q, D'Souza F, Willemsen PR, Donnelly GT, Donik C, Kocijan A, Jenko M, Jones LA, Guinaldo PC: **The potential of nano-structured silicon oxide type coatings deposited by PACVD for control of aquatic biofouling.** Biofouling 2009, **25:**55–67.

82. Martinelli E, Agostini S, Galli G, Chiellini E, Glisenti A, Pettitt ME, Callow ME, Callow JA, Graf K, Bartels FW: **Nanostructured films of amphiphilic fluorinated block copolymers for fouling release application.** Langmuir 2008, **24:**13138–13147.

83. Gudipati CS, Finlay JA, Callow JA, Callow ME, Wooley KL: **The antifouling and fouling-release perfomance of hyperbranched fluoropolymer (HBFP)-poly (ethylene glycol)(PEG) composite coatings evaluated by adsorption of biomacromolecules and the green fouling alga ulva.** Langmuir 2005, **21:**3044–3053.

84. Beigbeder A, Mincheva R, Pettitt ME, Callow ME, Callow JA, Claes M, Dubois P: **Marine fouling release silicone/carbon nanotube nanocomposite coatings: on the importance of the nanotube dispersion state.** J Nanosci Nanotechno 2010, **10:**2972–2978.

85. Beigbeder A, Labruyère C, Viville P, Pettitt ME, Callow ME, Callow JA, Bonnaud L, Lazzaroni R, Dubois P: **Surface and fouling-release properties of silicone/Organomodified Montmorillonite coatings.** J Adhes Sci Technol 2011, **25:**1689–1700.

86. Martinelli E, Sarvothaman MK, Galli G, Pettitt ME, Callow ME, Callow JA, Conlan SL, Clare AS, Sugiharto AB, Davies C, Williams D:**Poly(dimethyl siloxane) (PDMS) network blends of amphiphilic acrylic copolymers with poly(ethylene glycol)-fluoroalkyl side chains for fouling-release coatings. II. Laboratory assays and field immersion trials.** Biofouling 2012, **28:**571–582.

87. Genzer J, Efimenko K: **Recent developments in superhydrophobic surfaces and their relevance to marine fouling: a review.**Biofouling 2006, **22:**339–360.

88. Marmur A: **Super-hydrophobicity fundamentals: implications to biofouling prevention.** Biofouling 2006, **22:**107–115.

89. Wu A-F, Nakanishi K, Cho KL, Lamb R: **Diatom attachment inhibition: limiting surface accessibility through air entrapment.**Biointerphases 2013, **8:**1–10.

90. Bhushan B, Chae Jung Y, Koch K: **Micro-, nano- and hierarchical structures for superhydrophobicity, self-cleaning and low adhesion.** Philos T Roy Soc A 2009, **367:**1631–1672.

91. Zhang H, Lamb R, Lewis J: **Engineering nanoscale roughness on hydrophobic surface—preliminary assessment of fouling behaviour.** Sci Technol Adv Mat 2005, **6:**236–239.

92. Schenke-Layland K, Rofail F, Heydarkhan S, Gluck JM, Ingle NP, Angelis E, Choi C-H, MacLellan WR, Beygui RE, Shemin RJ, Heydarkhan-Hagvall S: **The use of three-dimensional nanostructures to instruct cells to produce extracellular matrix for regenerative medicine strategies.** Biomaterials 2009, **30:**4665–4675.

93. Warkiani ME, Bhagat AAS, Khoo BL, Han J, Lim CT, Gong HQ, Fane AG: **Isoporous micro/Nanoengineered membranes.** ACS Nano 2013, **7:**1882–1904.

94. Vucko MJ, Poole AJ, Carl C, Sexton BA, Glenn FL, Whalan S, de Nys R: **Using textured PDMS to prevent settlement and enhance release of marine fouling organisms.** Biofouling 2013, **30:**1–16.

95. Scardino AJ, de Nys R: **Mini review: biomimetic models and bioinspired surfaces for fouling control.** Biofouling 2010, **27:**73–86.

96. Decker JT, Kirschner CM, Long CJ, Finlay JA, Callow ME, Callow JA, Brennan AB: **Engineered antifouling microtopographies: an energetic model that predicts cell attachment.** Langmuir 2013, **29:**13023–13030.

Index